ENVIRONMENTAL REMEDIATION TECHNOLOGIES, REGULATIONS AND SAFETY

CARBON DIOXIDE EMISSIONS

NEW RESEARCH

ENVIRONMENTAL REMEDIATION TECHNOLOGIES, REGULATIONS AND SAFETY

Additional books in this series can be found on Nova's website under the Series tab.

Additional E-books in this series can be found on Nova's website under the E-book tab.

ENVIRONMENTAL REMEDIATION TECHNOLOGIES, REGULATIONS AND SAFETY

CARBON DIOXIDE EMISSIONS

NEW RESEARCH

MITCHELL CARPENTER
AND
EVERETT J. SHELTON
EDITORS

nova publishers
New York

For permission to use material from this book please contact us:
Telephone 631-231-7269; Fax 631-231-8175
Web Site: http://www.novapublishers.com

NOTICE TO THE READER

Additional color graphics may be available in the e-book version of this book.

Library of Congress Cataloging-in-Publication Data

Carbon dioxide emissions : new research / editors, Mitchell Carpenter and Everett J. Shelton.
pages cm
Includes bibliographical references and index.
ISBN 978-1-62257-436-0 (hardcover)
1. Carbon dioxide mitigation. I. Carpenter, Mitchell. II. Shelton, Everett J.
TD885.5.C3C373 2012
628.5'32--dc23

2012024289

Published by Nova Science Publishers, Inc. † New York

CONTENTS

PREFACE

This book examines new research in carbon dioxide emissions. CO2 is emitted in a number of ways naturally through the carbon cycle and through human activities like the burning of fossil fuels. Since the Industrial Revolution in the 1700s, human activities, such as the burning of oil, coal and gas, and deforestation, have increased CO2 concentrations in the atmosphere. Topics discussed in this compilation include mitigating climate change and CO2 emissions; CO2-e emissions from best practice landfills versus compost; evaluation of the potential carbon dioxide emissions reductions from the use of CHP systems in commercial buildings; renewable energy technology sources to mitigate CO2 emissions; and allocating carbon dioxide and other emissions with exergy.

Chapter 1 – A 2007 U.N. survey found that 54% of Americans advocate "wait-and-see" behavior on policies that mitigate climate change, i.e., they infer that climate mitigation actions can be deferred until there are clear signs of danger. By evaluating different cognitive factors that influence human behavior, this chapter proposes aframework that provides answers to an important question: why do people advocate wait-and-see behavior on climate change? One cognitive factor is misperceptions of feedback (i.e., ignorance of large feedback delays between CO_2 emission decisions and the corresponding changes in CO_2 concentration). Results reveal that the use of simulation tools, that provide repeated feedback about decision actions and corresponding consequences, is likely to enable people to overcome these misperceptions. A second factor is people's reliance on correlational or linear thinking (that the shape of CO_2 emissions and CO_2 concentration should look alike). Results reveal that the use of a physical representation (i.e., a picture of a problem in the form of a metaphor), simulation tools, and presenting problems such people's reliance on heuristics and biases enables them to make ecofriendly decisions (i.e., information presentation) is likely to enable people to overcome their correlational thinking. This chapter discusses concrete cognitive manipulations to improve public understanding about climate change and reduce the prevailing wait-and-see behavior.

Chapter 2 – Past research has shown that a majority of people exhibit robust linear thinking for nonlinear changes in their decision environment. The authors argue that linear thinking could be particularly problematic in the case of interpreting carbon-dioxide's (CO_2) lifetime in the earth's atmosphere. Participants from policy and non-policy backgrounds were asked to rank five ranges of CO_2 percentages to be removed from the atmosphere according to their impact on CO_2's lifetime in two separate conditions: Aid and no-Aid. In the Aid condition, participants were provided with a descriptive decision aid through instructions that might enable them to answer the problem correct, while this aid was absent in the no-Aid condition. Two problems were presented to each participant in random order: Linear, where a ranking based upon linear thinking yielded a correct rank order; and Nonlinear, where a ranking based upon linear thinking yielded an incorrect rank order. Results reveal that a majority of participants from both backgrounds responded linearly on both problems and although the decision aid had no effect on participants' correct responses, it enabled policy backgrounds to move away from responding according to linear thinking. The authors discuss implications of these findings on policymaking about climate change.

Chapter 3 – The steel industry is faced with a wide range of environmental concerns that are related to CO_2 emissions. Increasing demand to minimize the greenhouse gas emissions has intensified the pressure on steelmakers and significantly impacted the direction of the development of this industrial sector.

The mitigation of carbon dioxide emissions in the steel industry has been preformed by modifying and integrating currently available technolgies, especially more efficiently utilizing the heavily fossil-oriented energy sources. Although CO_2 emissions per ton steel produced has been noticeably reduced during the past decade, the steel industry still accounts for a significant portion of anthropogenic emissions of carbon dioxide and the modern steel plants operate close to the limits of what is technically possible.

Further reduction of carbon dioxide emissions is not possible without drastic technological changes. Here, the efforts of the steel industry on the development of new transformational technologies will be presented, especially introducing a novel flash ironmaking technology being developed at the University of Utah under the American Iron and Steel Institute's CO_2 Breakthrough Program. This technology will greatly reduce carbon dioxide emissions in the ironmaking step by eliminating coke oven and the pelletization or sintering step and by the use of hydrogen or natural gas as reductant/fuel. When it is operated with natural gas, the new technology will generate only 39% of carbon dioxide produced from the average blast furnace process.

Chapter 4 – Carbon dioxide (CO_2), methane (CH_4) and nitrous oxide (N_2O) are together major greenhouse gas emissions from waste management practices. Methane and nitrous oxide are predicted to cause as much global warming in terms of CO_2 equivalents (CO_2e) as carbon dioxide over the next 20 years, according to the Intergovernmental Panel on Climate Change. Methane's global warming impact (GWI) over 100 years is 25 times that of CO_2. This impact is higher in the first 20 years. N_2O is even more potent, with 298 times the GWI of carbon dioxide, based findings by the Intergovernmental Panel on Climate Change (IPCC 2007).

It is well-known that methane is produced in anaerobic digestion, as in landfill. It is less well-known that methane and N_2O are both emitted in small but significant quantities when organic material is composted (IPCC 2006).

The City of Melbourne, Australia has strict Environment Protection Agency regulations controlling landfills. Best practice landfills have measures such as vacuum gas extraction which can capture greater than 75% of the methane produced (NGERS Technical Guidelines 2010), and this methane is used in the generation of electricity, thus reducing CO_2 emissions from coal-fired energy generation. Yet Melbourne also has a policy of diversion of municipal waste to compost, which reduces this yield of landfill gas.

The State of Victoria set a target for diverting wastes from landfill by 65%, between 2005 and 2014. Victoria's "Towards Zero Waste Strategy Report 2010" states that this policy "reduces energy use and carbon emissions and helps mitigate climate change". Is this true? This study compares expected CO_2-e emissions from landfill (with methane capture) versus emissions from municipal composting, using local data on Melbourne's waste composition. Calculations were based on Australian guidelines for estimating CO_2-e emissions (DCCEEE 2011) and IPCC (2006).

Findings: Cumulative CO_2-e emissions in Victoria are in fact significantly lower when organic material is sent to best practice landfill sites, rather than to well-turned municipal compost sites.

Chapter 5 – Combined heat and power (CHP) systems generate electricity near the site of use and capture the waste heat for space heating or hot water. CHP systems have the potential to reduce the CO_2 emissions resulting from power production when compared to a reference system where all the power is purchased from the electricity grid. However, the potential to reduce emissions with the use of CHP systems depends on the electrical and thermal requirements of the building. This chapter presents a model to evaluate the potential emission reductions resulting from the implementation of CHP systems. The model is based on two main parameters. The first parameter, R_e, is the percentage of the average hourly electricity demand of the building satisfied by the CHP system, and the second parameter, R_h, is the fraction of the thermal load that is satisfied by the CHP system. In order to demonstrate how to apply the model, four different type of buildings located in Chicago, Illinois are evaluated for their potential to reduce CO_2 emissions using a CHP system.

Chapter 6 – Owing to global energy crisis, energy conservation through energy efficient buildings has acquired prime importance. Hence, this chapter deals with various renewable energy technologies that can be integrated with building. The three main aspects which should be given consideration for energy saving in a building include first and foremost the building design, secondly the low energy building materials in building construction and lastly the use of renewable energy technologies for various applications. The solar passive heating/cooling design features like sunspace, Trombe walls, earth to air heat exchangers, adobe house, day-light etc can lead to considerable energy saving at the stage of building design. Further, low embodied energy building materials should be chosen like fly ash bricks; fiber reinforced bricks; and stabilized mud blocks etc. which reduce energy consumption at building construction stage. The use of such low embodied energy materials has gained importance nowadays. The energy requirements of building can be cut down to minimum by use of renewable energy technologies like solar photovoltaic, wind turbines, solar thermal devices etc. These technologies are discussed in this chapter with their economic analysis and environmental impacts. In the present alarming situation of global warming, it is known that about 1 kg of CO_2 gets emitted in to the environment corresponding to generation of 1 kWh unit of electrical energy from the coal thermal power plants. Hence, for sustainable future of the world, there is urgent need for use of renewable energy sources to mitigate CO_2 emissions. Today, alternative renewable energy technologies are becoming prime important to meet at least 10 % of the total building energy requirements in European countries like U.K.

Chapter 7 – The allocation of carbon dioxide emissions from cogeneration is often challenging, as existing methods are not soundly based, inconsistent, overly complex and difficult to apply. The author proposes that exergy methods can form the basis for rationally and meaningfully allocating cogeneration-based carbon dioxide emissions, and describes and illustrates such methods in this chapter and compares them with other methods. The results suggest that the exergy-based approach provides a sensible basis for a meaningful overall approach for carbon dioxide emissions trading. It is concluded that the exergy-based method for allocating carbon dioxide emissions from cogeneration is rational, useful and superior to other methods. The results aid designers of energy systems, and decision and policy makers in companies and government, especially regarding how emissions should be reduced, how and where cogeneration should be used, and how such carbon dioxide emissions can be traded.

In: Carbon Dioxide Emissions: New Research
Editors: M. Carpenter and E. J. Shelton
ISBN: 978-1-62257-436-0

Chapter 1

WHY DO WE DEFER ACTIONS ON CLIMATE CHANGE? A COGNITIVE PERSPECTIVE

Varun Dutt *

School of Computing and Electrical Engineering, School of Humanities and Social Sciences, Indian Institute of Technology, Mandi, H.P., India
Dynamic Decision Making Laboratory, Department of Social and Decision Sciences, Carnegie Mellon University, Pittsburgh, PA, US

ABSTRACT

A 2007 U.N. survey found that 54% of Americans advocate "wait-and-see" behavior on policies that mitigate climate change, i.e., they infer that climate mitigation actions can be deferred until there are clear signs of danger. By evaluating different cognitive factors that influence human behavior, this chapter proposes aframework that provides answers to an important question: why do people advocate wait-and-see behavior on climate change? One cognitive factor is misperceptions of feedback (i.e., ignorance of large feedback delays between CO_2 emission decisions and the corresponding changes in CO_2 concentration). Results reveal that the use of simulation tools, that provide repeated feedback about decision actions and corresponding consequences, is likely to enable people to overcome these misperceptions. A second factor is people's reliance on correlational or linear thinking (that the shape of CO_2 emissions and CO_2 concentration should look alike). Results reveal that the use of a physical representation (i.e., a picture of a problem in the form of a metaphor), simulation tools, and presenting problems such people's reliance on heuristics and biases enables them to make ecofriendly decisions (i.e., information presentation) is likely to enable people to overcome their correlational thinking. This chapter discusses concrete cognitive manipulations to improve public understanding about climate change and reduce the prevailing wait-and-see behavior.

* Correspondence concerning this article should be addressed to Prof. Varun Dutt, Ph.D., School of Computing and Electrical Engineering, School of Humanities and Social Sciences, Indian Institute of Technology, Mandi, PWD Rest House 2nd Floor, Mandi - 175001, H.P., India. E-mail: varundutt@iitmandi.ac.in.

Keywords: Wait-and-see behavior; climate change; linear thinking; misperceptions of feedback; simulation tools; physical representation; information presentation

INTRODUCTION

Despite strong scientific consensus about causes and risks of climate change, the general public seems to exhibit a complacent attitude towards actions that benefit Earth's climate (Bostrom et al., 1994; Leiserowitz, 2007; Read et al., 1994; Weber, 2006). Recent surveys have shown that most Americans exhibit "wait-and-see" behavior, according to which they infer that mitigation actions on climate change (e.g., reductions in yearly greenhouse gas (GHG) emissions) can be deferred until there is greater evidence that climate change is harmful (Sterman and Booth Sweeney, 2002, 2007; Leiserowitz, 2007). People's wait-and-see behavior on climate policies is currently widespread. For example, a 2007 U.N. survey found a large majority of U.S. respondents (54%) advocating a wait-and-see or go-slow approach to emission reductions (Sterman, 2008; Leiserowitz, 2007), and larger majorities favoring the wait-and-see option in Russia, China, and India (Sterman, 2008; Leiserowitz, 2007).

Moreover, recent evidence shows that wait-and-see behavior is also widespread among policymakers who directly decide on policies that mitigate climate change. For example, Tony Abbott, opposition frontbencher and a senior member of the Liberal Party of Australia in response to a news-poll commented, "The government should not be rushing headlong into any premature trading scheme [for policies that mitigate climate change]" (Conway, 2009, para. 2). Also, the former White House science adviser John Marburger briefed a Senate Panel on climate change, saying, "We know we have to make very large changes *if* this [climate] turns out to be a problem. The consequences of human-induced global warming could be quite severe" (Jones, 2002, para. 2). At the same briefing, the U.S. administration stood behind its wait-and-see behavior by claiming that, "we should only slow the growth of greenhouse gas emissions (GHGs), and - as the science justifies - stop, and then reverse that growth" (Jones, 2002, para. 3). Similarly, Fred Singer, professor emeritus environment sciences, University of Virginia, expressed a stronger wait-and-see view by commenting that, "human activities are not influencing the global climate in a perceptible way. Climate will continue to change, as it always has in the past, warming and cooling on different time scales and for different reasons, regardless of any human action" (Singer, 2009, p. 1).

Amidst these wait-and-see statements, there is a growing consensus that Earth's climate is likely changing as a direct result of human actions such as burning of fossil fuels and deforestation that lead to greenhouse gas emissions (IPCC, 2007a; Arctic Climate Impact Assessment, 2004). Although there are serious concerns for Earth's climate among certain climate scientists, citizens, policymakers, and governmental officials, it is clear that a majority of people currently prefer to exhibit a risk-seeking wait-and-see behavior on initiating any actions that mitigate climate change.[1] People's wait-and-see behavior is risk-seeking and problematic because holding off actions that mitigate climate change is likely to

[1] Defined more formally, consider these two choices:A. Paying a certain amount now for sure and reducing the future impacts of climate change; or, B. Defer paying now and paying a larger but an uncertain amount in the future as a result of climate change.Then, a choice for option B is a measure of risk-seeking behavior on climate change. This risk-seeking choice also exhibits wait-and-see behavior to climate change mitigation actions in the status quo.

invite climate "black swans" in the future, i.e., high-impact, hard-to-predict, and rare climate consequences beyond the realm of normal expectations (Taleb, 2007). That is because, unlike simple systems that have short delays between detection of a problem and the implementation of corrective actions (e.g., boiling beans in a cooker where upon hearing the whistle one could immediately remove the cooker from the flame), for a complex system like Earth's climate there are long delays between the decision to mitigate emissions and the corresponding changes in atmospheric greenhouse gas (GHG) concentrations (IPCC, 2007b; Sterman and Booth Sweeney, 2002, 2007; Sterman, 2008). Therefore, for the climate change problem, one cannot afford to delay acting (i.e., exhibit wait-and-see behavior) till the last day and then decide to take actions when there are clear signs of danger due to climate change. On account of deferring immediate mitigation actions and the long feedback delays in the climate system, the climate black swans experienced in the future are likely to be negative consequences in the form of severe storms, sudden melting of polar ice-caps and droughts that occur in different parts of the world. Also, the cost of adaptation to these negative consequences is estimated to be very large (IPCC, 2007a; Nordhaus, 1994).

Currently, there is lot of technological and engineering research on developing mitigation options for climate change. Some of these technological options include carbon capture and sequestration (CCS), geo-engineering, and switching to renewable sources of energy (IPCC, 2007a). This technological and engineering research is very important and very much needed. However, as the decisions to adopt different technological and engineering options ultimately rests with people, therefore, research that studies the role of people, their understanding of climate change, and their wait-and-see behavior on climate change is also critically needed, though currently lacking in the literature (APA, 2009).

This chapter attempts to fill this void and directly considers the role of people, their wait-and-see behavior, and their understanding of climate change. In order to explain people's wait-and-see behavior, this chapter takes an approach that is different from applying standard economic methods like cost-benefit analysis, where one would justify people's wait-and-see behavior in terms of sure costs now versus larger and uncertain costs of consequences that are distant in the future and that are discounted at high discount rates (Monbiot, 2009; Nordhaus, 2008; Stern, 2006).[2] Rather, this chapter proposes a framework of people's wait-and-see behavior by considering different cognitive factors that are peculiar to how humans make day-to-day decisions. Therefore, here I consider the human perceptions and decisions on climate change from a cognitive perspective. Given that climate change is already occurring in some parts of the world while people continue to exhibit wait-and-see behavior on the problem (Arctic Climate Impact Assessment, 2004), the main purpose of this chapter is to discuss from a cognitive perspective the following research questions:

- Q1: Why do people want to defer mitigation actions on climate change when science currently shows signs of a rapid change in Earth's climate?

---

[2] Discount rate is the rate at which a future cost (F) is discounted to the present cost (P). The exact relationship is $F = P * (1 + i/100)^n$, where i is the per annum discount rate and n is the number of years in the future after which cost F is incurred. If future costs of climate consequences is large, then even for a moderate discount rate (~ 8%), the calculated present costs will be very small. From an economic perspective, these costs of future consequences in the status quo will be small and this observation is likely to make people exhibit wait-and-see behavior on policies that mitigate climate change.

- Q2: Furthermore, is there a way to influence people's wait-and-see behavior through certain manipulations?

I discuss both questions by citing some recent laboratory research. This research involves systematically manipulating certain conditions and collecting behavioral data, where human participants make judgment and choices on climate problems in a controlled laboratory environment.

BACKGROUND AND CONTRIBUTIONS

There could be several potential reasons for people's wait-and-see behavior on climate change. Recent research has concentrated on the effects of the current costs of mitigation actions (using a high discount rate for future costs) and denial, i.e., ignorance of the climate change problem (Monbiot, 2009; Nordhaus, 2008; Stern, 2006). These economic reasons on cost and benefits have been the topic of many current investigations.

Furthermore, there is also research that has considered the role of motivational factors on environmental decisions (e.g., political ideology, perceptions of needs versus luxuries, and core psychological needs etc.) (APA, 2009; Hardisty, Johnson, and Weber, 2010). For example, in November, 2008 there were devastating fires in southern California disrupting the ecology of that place; however, very few people (including those that had lost houses in the previous fires) decided to move out of the region. One motivational factor that explains this behavior could be place attachment (Gifford, 2007), i.e., continued attachment to family, job, and community, a motivation that can be more salient in the aftermath of adverse events, when fears have faded, than the goal of avoiding a low-probability future disaster. More recently, however, in other but similar situations involving organ donations (Johnson and Goldstein, 2003) and insurance decisions (Johnson, Hershey, Meszaros, and Kunreuther, 1993), the role of cognitive factors like status quo bias and inertia have been documented as reasons for environmental factors rather than the motivational factors. As there is evidence in favor of both cognitive and motivational factors, they are both likely to influence people's decisions on environmental problems.

In this regard, although the role of motivational and economic factors in connection to environmental problems is documented in literature, the role of cognitive factors has been somewhat less studied and downplayed (APA, 2009). Due to the inseparable interaction of humans' actions on climate, it is important to discuss the role of certain cognitive factors that are likely to influence people's wait-and-see behavior on climate change. Detailed below are some of the cognitive factors and how this chapter builds upon these factors to suggest a framework of people's wait-and-see behavior.

MISPERCEPTIONS OF FEEDBACK

According to the misperceptions of feedback (MOF) hypothesis (Sterman 1989), people ignore the feedback delays present in their decisions in a dynamic system. In the case of the Earth's climate, the MOF hypothesis suggests that people are likely to fail to take into account the long time delays between increases in carbon-dioxide (CO_2) emissions and the subsequent increases in CO_2 concentration, and the delays between increases in CO_2 concentration and its effects on increasing average atmospheric temperature (Sterman and Booth Sweeney 2002; 2007). In Earth's climate, an increase in CO_2 emissions does not increase CO_2 concentration and atmospheric temperature immediately, but after a long delay where it might be too late to act to avoid significant impact of these increases.

Moxnes and Saysel (2009) have built on Sterman and Booth Sweeney's (2007) results by focusing on how people incorporate feedback delays in their decisions and regulate CO_2 emissions in order to reach an attainable CO_2 concentration goal in a simulated climate system. They tested participants' ability to control CO_2 concentration to 300 GtC above the pre-industrial level in a period between the years 2000 and 2100, where participants made emission decisions every 10 years. They tested participants in different conditions that mimicked the working of the climate system, with repeated feedback about decisions and the resulting changes in CO_2 concentration. Participants entered ten numbers which represented their emission decisions every ten years over a 100 year period. In all but one feedback condition, participants entered *all* ten emission values at one time and then observed the effects of their decisions on the CO_2 concentration over the 100 year period. In conditions without feedback, Moxnes and Saysel's (2009) results coincided with results from Sterman and Sweeney (2007): Participants showed a general tendency to overshoot the goal level, misperceiving feedback delays in their emission decisions. In the feedback condition, however, Moxnes and Saysel (2009) gave participants the ability to make repeated emission decisions every 10 years and to observe the effects of these decisions on the CO_2 concentration before the next 10 yearly emission decision was made. Within a 10 year period, the emissions remained constant at values which were set at the start of the period. Results show that providing repeated feedback helped participants change their strategy over time, and may have helped them to reduce their misperceptions of feedback delays.

Dutt and Gonzalez (2012a) extend results of Moxnes and Saysel (2009) and Sterman and Booth Sweeney (2002; 2007) in a task called the Dynamic Climate Change Simulation (DCCS),[3] where participants control CO_2 concentration to a goal level under different kinds of feedback delays in inputs (CO_2 emissions) and outputs (CO_2 absorptions). The design of DCCS has been motivated from a generic dynamics stocks and flows task (Gonzalez and Dutt, 2011) and the use of the "bathtub metaphor," where CO_2 emission is represented by a tap, CO_2 absorption is represented by a drain, CO_2 concentration is represented by the amount of liquid in the tub, and the tub itself represents Earth's atmosphere.

Using DCCS in an laboratory experiment, Dutt and Gonzalez (2012a) have systematically varied two types of feedback delays in DCCS, the frequency of making emission decisions (frequently or infrequently) and rates of CO_2 absorption (rapid or slow). They show that when laypeople are presented with the problem of controlling carbon-dioxide (CO_2) concentration in DCCS, people who perform in realistic climate conditions (where the rate of CO_2 absorptionis slow and emission decisions are made infrequently) produce larger CO_2 emissions compared to people who perform in more optimistic conditions (where the rate of CO_2 absorption are fast and emission decisions are made frequently). These results are explained by the difficulty individuals have in perceiving feedback delays in dynamic systems

[3] The DCCS simulation can be downloaded for free under an academic license from: http://downloads.ddmlab.com/?action=formandpackage_id=2.

(Diehl and Sterman, 1995; Booth Sweeney and Sterman, 2000; Sterman and Booth Sweeney, 2007; Sterman, 1989; Sterman, 2000; Sterman, 2002; Sterman, 2008). According to Dutt and Gonzalez (2012a), people in conditions of feedback delay exhibit wait-and-behavior in their yearly CO_2 emission policies because people continue to maintain their emissions at a high value (as is the case in the world presently) and when they want to cut the emission to meet absorption upon hitting the goal (which is needed) it becomes too late for them to act. Here, cutting down emission to meet absorption is something that opposes people's wait-and-see behavior and is needed right from the start in the task, rather than only upon reaching the goal. However, results also show that participants do learn to improve their emission decisions by cutting down emissions over many repeated trials of decision making in DCCS. Thus, one manipulation for reducing people's misperceptions of feedback delays is the use simulation tools like DCCS that compress time and space and that improve people's understanding of the dynamics of Earth's climate over repeated interactions with these tools.

RELIANCE ON CORRELATION HEURISTIC

In addition to the human misperceptions of feedback delays (Dutt and Gonzalez, 2012a; Sterman, 1989), a less explored cognitive factor for people's wait-and-see behavior is the human tendency to rely on "correlation heuristic" while making judgments about a level or accumulation (e.g., CO_2 concentration in the atmosphere) based on CO_2 emission (inflow) and absorption (outflow), respectively (Cronin and Gonzalez, 2007; Cronin, Gonzalez, and Sterman, 2009).[4] According to correlation heuristic, people incorrectly assume that the shapes of accumulation and inflow overtime should be identical. For example, it has been shown that highly educated people rely on correlational thinking, assuming that if the task demands one to increase the GHG concentration to stabilize at a higher level than the status quo, then GHG emissions should also increase and stabilize in an identical manner to the GHG concentration (Sterman and Booth Sweeney, 2002; Sterman and Booth Sweeney, 2007). People's reliance on correlation heuristic for Earth's climate supports wait-and-see behavior and violates basic laws of physics. That is because, for Earth's climate, the CO_2 concentration has been increasing nonlinearly over time (years) due to an approximate linear increase in CO_2 emissions every year (IPCC, 2007). By relying on correlation heuristic, people are likely to judge the nonlinear CO_2 concentration's shape to be linearly increasing over time, i.e., similar to the linearly increasing shape of CO_2 emissions. Consequently, such linear judgments are likely to make people underestimate the actual nonlinear increase in CO_2 concentration, undermine the seriousness of the climate problem, and make them exhibit wait-and-see behavior.

As reliance on correlation heuristic is likely to be very problematic for immediate actions on climate change, in this chapter, I discuss a number of manipulations that aim to reduce people's reliance on this heuristic and the associated misconceptions. One of the manipulations relies on using DCCS among people from both the science and mathematics backgrounds (STEM) as well as the humanities and arts backgrounds (non-STEM) (Dutt and

[4] Inflow and outflow are rates of change and the change in accumulation is the integral of the difference between the inflow and outflow.
Therefore, $Accumulation\ (t) = \int_1^t [Inflow(t) - outflow(t)].dt + Accumulation\ (0)$.

Gonzalez, 2012a). Dutt and Gonzalez (2012b) put participants from both STEM and non-STEM backgrounds in two separate conditions in a laboratory experiment. In one (control) condition, participants sketched the CO_2 emission and absorption trajectories they thought to correspond to a given CO_2 concentration stabilization (CS) scenario over 100 years on a sheet of paper (i.e., the CS Task, Sterman, 2008). In a separate (experimental) condition, participants controlled the CO_2 concentration as close as possible to a CO_2 stabilization trajectory (same as that given in CS task) by indirectly manipulating the CO_2 emissions and absorptions in DCCS over 100 years. Participants' performance in DCCS was followed by their performance in the CS task. Results revealed that the DCCS manipulation worked effectively in reducing participants' reliance on correlation heuristic among both STEMs and non-STEMs in the CS task after performing in DCCS compared to when participants performed in the CS task alone. However, the benefits of DCCS performance were greater for STEMs compared to non-STEMs: the reduction in reliance on correlation heuristic was greater for STEMs compared to non-STEMs in the CS task after performing in the DCCS. One explanation of this observation is that previous exposure to mass balance and energy balance concepts in mathematics, science, and engineering enables STEMs to improve their sketches in the CS task because they focus less on the surface (i.e., irrelevant) features of decision problems, and focus more on the more fundamental underlying structure of the problem (Chi, Feltovich, and Glaser, 1981; Gonzalez and Wong, 2012).

Furthermore, Dutt and Gonzalez (in press) hypothesize that people's reliance on correlation heuristic and their consequent underestimation of nonlinear accumulation are influenced by the format in which these problems are presented to them. Here, Dutt and Gonzalez (in press) motivate and investigate the use of a physical representation to communicate climate problems as an alternative to other conventionally used representations that include text descriptions and mathematical graphs. Physical representation presents a climate problem using a picture that works as a "metaphor." For example, consider a mathematical problem where the length (L) of a square's side is doubled and one needs to calculate the new area. A physical representation of the problem will be one where a single square tile of side L is replaced by four square tiles, each of side L (i.e., to form a square of side 2L) in a picture. In a series of experiments, Dutt and Gonzalez (in press) have shown that the physical representation reduces people's reliance on correlation heuristic and their underestimation of CO_2 accumulation when the representation's use is compared to other graphical and text representations in different problems and contexts (that are even different from the climate context).

Finally, Dutt and Gonzalez (2012c (this book); 2012d) discuss an information presentation manipulation that is hypothesized to enable people to improve their decisions by directly relying on the correlation heuristic rather than reducing this reliance. According to the information presentation manipulation, one method of improving people's decisions is to present ecofriendly options in such a manner that these options appear more attractive to people when they rely on linear thinking. Therefore, if people choose an ecofriendly option by relying on linear thinking, then they will pick an option that mitigates climate change without any alteration in their incorrect linear thought process. For example, Dutt and Gonzalez (2012c) discuss how linear thinking could be particularly problematic in the case of interpreting carbon-dioxide's (CO_2) lifetime in the Earth's atmosphere. They took participants from policy and non-policy backgrounds and asked them to rank five ranges of CO_2 percentages to be removed from the atmosphere according to their impact on CO_2's lifetime in two separate conditions: Aid and no-Aid. In the Aid condition, participants were provided with a descriptive decision aid through instructions that would likely enable them to answer the problem correctly, while this aid was absent in the no-Aid condition. Two problems were presented to each participant in random order: Linear, where a ranking based upon linear thinking yielded a correct rank order; and Nonlinear, where a ranking based upon linear thinking yielded an incorrect rank order. Results revealed that a majority of participants from both backgrounds responded linearly on both problems and got the problem correct where a linear response gave the correct answer. Furthermore, although the decision aid had no effect on participants' correct responses, it enabled policy backgrounds to move away from responding according to linear thinking. In another example, Dutt and Gonzalez (2012d) have applied the same information presentation manipulation to ecological (eco) taxes, which are promising mechanisms to enable ecofriendly decisions; however, which do not currently enjoy popular public support. Dutt and Gonzalez (2012d), show that thinking linearly people rely on the linear-thinking heuristic, i.e., they prefer a small price increase while associate larger price increases to mean proportionally greater benefits or improvements in quality. Participants were asked to choose between two eco-tax increases in two decision problems: in one, the smaller eco-tax increase resulted in greater CO_2 emissions reduction, while in the other, the smaller increase resulted in lesser reduction. Although larger eco-tax increases did not always save more CO_2 emissions, a majority of participants preferred the smaller eco-tax increases, while judging larger tax increases to cause greater reductions in CO_2 emissions (i.e., they relied on the linear-thinking heuristic). Since participants rely on the linear-thinking heuristic in reaching their preferences and judgments about eco-taxes, eco-tax policies are likely to benefit by presenting information such that smaller tax increases (which are linearly more attractive and are likely to be chosen by people) cause greater CO_2 emissions reductions.

In the different manipulations discussed above, the approach taken is either to help people reduce their reliance on the discussed cognitive factors, or to use that very reliance to promote decisions that mitigates climate change. In summary, this chapter theoretically motivates certain cognitive factors and manipulations that affect and influence people's wait-and-see behavior in a series of laboratory experiments. In the next section, I bring together several cognitive factors into a framework to explain people's wait-and-see behavior.

A Framework for Wait-and-See Behavior on Climate Change

Although prior research has expounded economic reasons for people's wait-and-see behavior on climate change, currently very little is known about how certain cognitive factors might influence this behavior. Considering the global nature of the climate problem and the associated large negative consequences in the future (IPCC, 2007a; 2007b; Sterman, 2008), the research discussed here contributes to certain cognitive factors that are likely to influence people's wait-and-see behavior. This contribution is significant because no matter what technology or policy solutions are worked out for averting future climate change, these solutions will suffer from the problem of adoption and acceptance by the common and global populace without a proper understanding of how humans are likely to respond to these solutions and whether they are going to defer acting on these solutions (Sterman, 2008; APA, 2009).

Figure 1 presents a framework about people's wait-and-see behavior by combining the role of different cognitive factors (described above) to influence this behavior. As shown in the figure, there are a number of manipulations proposed (discussed above) that influence people's wait-and-see behavior through their cognitive factors.

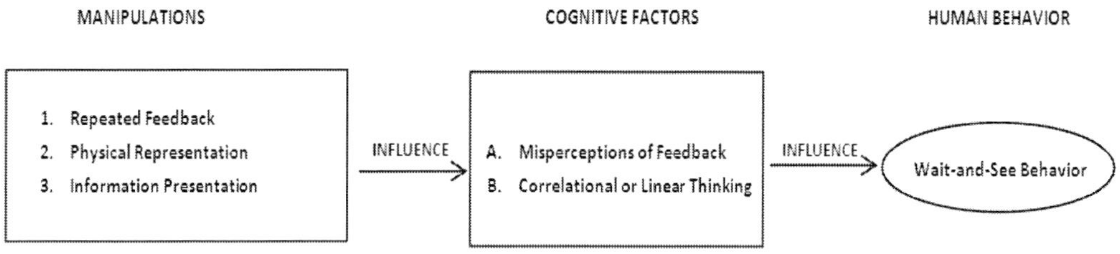

Figure 1. A framework for wait-and-see behavior on climate change as a consequence of a number of cognitive factors. Certain manipulations indirectly affect people's wait-and-see through their cognitive factors.

The cognitive factors include: people's linear perception of GHG accumulations in climate problems and their misperceptions of feedback delays present in different actions and consequences in these problems. Thus, while designing technology or policy solutions for averting climate change, policymakers need to be careful and pay attention to the proposed cognitive factors that are likely to influence people's wait-and-see judgments and choices on these solutions. Furthermore, as shown in Figure 1, there are a set of manipulations that indirectly influence people's wait-and-see behavior on climate change. Their influence is mediated through cognitive factors. One of these manipulations is to provide repeated feedback in simulation tools and in the real world after people make judgment and choices. As discussed above, repeated feedback enables people to understand and observe the consequences of their decision actions and consequently correct their future decision actions. Another manipulation is to use the physical representation, i.e., to exhibit climate problems using pictures as metaphors that enable people to improve their understanding, reduce their misconceptions, and improve their decision judgments in these problems.

However, many times it might be difficult to change or alter people's reliance on cognitive factors, improve their decision making, and influence their wait-and-see behavior. Therefore, rather than change people's existing thinking processes, often times it might be beneficial to exploit these processes to enable people to improve their decisions. As discussed above, according to the information presentation manipulation, presenting options that are better for Earth's climate as options that people are likely to choose due to their cognitive factors, will enable people to improve their decisions on climate change without any alteration in the way they decide on day-to-day problems.

Moreover, the proposed framework also contributes to furthering the study of cognitive factors. This framework extends what is currently known on how human behavior is impacted by different cognitive factors, namely, misperceptions of feedback and correlational or linear thinking. For example, currently little is known about how certain manipulations are likely to reduce people's reliance on correlational or linear thinking. Here, I discuss the use of repeated feedback and physical representation as two promising manipulations for reducing people's reliance on linear thinking.

CONCLUSION

In this chapter, I proposed a framework of people's wait-and-see behavior on climate change due to the interplay of a number of cognitive factors that are peculiar to human beings. Results show that cognitive factors like misperceptions of feedback and correlational or linear thinking influence people's wait-and-see behavior about policies that mitigate climate change. For example, due to misperceptions of feedback, people ignore the feedback delays between climate mitigation actions and their associated consequences and this ignorance results in wait-and-see behavior. Moreover, thinking linearly, people tend to underestimate the actual nonlinear accumulation of greenhouse gases in Earth's atmosphere. This underestimation is likely to result in people undervaluing the urgency of the climate change problem thus resulting in wait-and-see behavior.

There is little doubt that climate change will become a serious problem for society if we decide to postpone mitigation actions to a time in the future (IPCC, 2007b; Sterman, 2008). Therefore, in order to make people act on climate change in the status quo, we must propose education policies that correct people's misperceptions about climate processes. According to results discussed in this chapter, repeated feedback in simulation tools (like DCCS) about the consequences of setting different CO_2 emission policies helps people to improve their control of CO_2 concentration over many trials. Thus, both during policymaking on climate and while imparting climate education in schools, it would be desirable to incorporate the use of simulation tools that supplement normal methods of policymaking and education. In this regard, the repeated feedback manipulation is also likely to help people to make energy efficient decisions that save electrical energy in their households. For example, imagine an air conditioner with a meter display that tells people how much time it has been running and how much energy and money it has accumulated since the first day of the month. Such an intervention that provides repeated feedback about energy and money used is likely to make people conscious of the cause-and-effect relationship between their decision to use the air conditioner and the corresponding high cost consequences in their monthly electric bill. As people are likely to avoid high costs, the end result is likely to be an efficient use of the air conditioner that saves energy.

Policies that supplement the use of repeated feedback in simulation tools with normal methods are good, but these policies also need to be sensitive to the differential needs of people with and without backgrounds and education in science and mathematics. In this regard, results suggests further research that investigates the design of certain background-specific features as part of these tools which when incorporated are likely to benefit people with science and non-science backgrounds equally.

Furthermore, results also suggest the use of physical representation (which represents a problem using a picture that acts as a metaphor) as an effective method of representing different climate and non-climate problems compared to text and mathematical graphs. For example, a physical representation that shows the energy inflow and outflow in the U.S. using a picture is likely to be very useful to policymakers in order to target specific sectors of the economy (e.g., industrial or transportation) for energy efficiency.

Similarly, physical representation is also likely to succeed in conserving Earth's natural resources. For example, in 2008, the City of New York and partners launched an advertising campaign to promote recycling awareness in the city. The campaign used physical representation involving metaphor and analogy by comparing the huge amount of recyclable paper thrown away in New York City annually to be equivalent to filling the entire Empire State Building. The advertisement created a "picture" of the iconic skyscraper composed entirely of discarded magazines and catalogs (i.e., a physical representation), and this metaphor enabled people to improve their recycling behavior.

Given these successes, the physical representation manipulation also seems promising considering the fact that unlike a simulation tool that needs a computer machine, the physical representation does not require a computer to be projected, rather it could be drawn on a piece of paper (as the New York City advertisement campaign does).

Thus, the physical representation could be directly used in climate reports and newspapers without any need of computer technology to accompany these publications. Consequently, it is likely that the physical representation has a greater societal reach considering that in many parts of the developing and underdeveloped world, the advent of computer technology and internet is still a distant dream, and in these places, people gather most of their news from newspapers and other printed material.

Finally, results suggest that often times it might be difficult to change people's reliance on cognitive factors and here aligning problems with people's reliance on these factors (like linear thinking) is likely to enable a majority of them to make ecofriendly decisions. Such interventions, that target and use people's reliance on cognitive factors to improve their decision making, are likely to enable proliferation of ecofriendly policies (e.g., eco-taxes) that benefit society and that are less problematic in terms of gathering public support (as they don't change the way people think). Also, the power of this information presentation manipulation is that the same manipulation can be used for many different societal problems that the world faces day-to-day: pollution in river and seas, cigarette smoking, and throwing garbage in public places etc.

ACKNOWLEDGMENTS

This research is supported by a fellowship from the Steinbrenner Institute, Carnegie Mellon University to Prof. Varun Dutt while he was persuing his Ph. The author would like to thank the computing resources provided by Dynamic Decision Laboratory, Carnegie Mellon University and Indian Institute of Technology, Mandi. The author would like to thank Dr. CleotildeGonzalez, Dr. Mitchell J. Small, Dr. Lorrie F. Cranor, and Dr. Elke U. Weber for their help in reviewing this manuscript and providing insightful comments.

REFERENCES

APA. (2009). *Psychology and Global Climate Change: Addressing a Multi-faceted Phenomenon and Set of Challenges*. Retrieved September 7, 2009, from American Cognitive Association Web site: http://www.apa.org/ science/climate-change/.

Arctic Climate Impact Assessment.(2004). Arctic Climate Impact Assessment statement by Dr. Robert W. Corell, Chair Arctic Climate Impact Assessment. Retrieved on 25th January, 2012 from: http://www.acia.uaf. edu/PDFs/Testimony.pdf.

Booth Sweeney, L., and Sterman, J. D. (2000). Bathtub dynamics: Initial results of a systems thinking inventory. *System Dynamics Review,16*(4), 249-286.

Bostrom, A., Morgan, M. G., Fischhoff, B., & Read, D. (1994). What do people know about global climate change? 1. Mental models. *Risk Analysis*, *14*(6), 959-997.

Chi, M. T. H., Feltovitch, P. J., & Glaser, R. (1981). Categorization and representation of physics problems by experts and novices. *Cognitive Science, 5*, 121-152.

Conway, D. (2009). *Wait-and-see on climate change: Abbott*. Retrieved November 15, 2009, from The Age Newspaper Web site: http://news. theage.com.au/breaking-news-national/wait-and-see-on-climate-change-abbott-20090728-dztr.html.

Cronin, M., & Gonzalez, C. (2007). Understanding the building blocks of system dynamics. *System Dynamics Review, 23*(1), 1-17.

Cronin, M., Gonzalez, C., & Sterman, J. D. (2009). Why don't well-educated adults understand accumulation? A challenge to researchers, educators and citizens. *Organizational Behavior and Human Decision Processes, 108*(1), 116-130.

Diehl, E., & Sterman, J. D. (1995). Effects of feedback complexity on dynamic decision making. *Organizational Behavior and Human Decision Processes*, *62*(2), 198-215.

Dutt, V., & Gonzalez, C. (2012a). Human Control of Climate. *Climatic Change, 111*(3), 497-518. doi: 10.1007/s10584-011-0202-x.

Dutt, V., & Gonzalez, C. (2012b). Decisions from Experience reduce misconceptions about Climate Change. *Journal of Environmental Psychology*, *32*(1), 19-29.

Dutt, V., & Gonzalez, C. (2012c). Responding linearly in nonlinear problems: Application to earth's climate. In M. Carpenter & E. J. Shelton (Eds.), *Carbon Dioxide Emissions: New Research* (pp. xx-yy). Hauppauge, New York, USA: Nova Publishers.

Dutt, V., & Gonzalez, C. (2012d). Enabling eco-friendly choices by relying on the proportional-thinking heuristic. Under review.

Dutt, V., & Gonzalez, C. (in press). Reducing the Linear Perception of Non-Linearity: Use of a Physical Representation. *Journal of Behavioral Decision Making*. doi: 10.1002/ bdm.759.

Gifford, R. (2007). *Environmental psychology: Principles and practice*. Coleville, WA: Optimal Books.

Gonzalez, C., & Dutt, V. (2011). A Generic Dynamic Control Task for Behavioral Research and Education. *Computers in Human Behavior*, *27*, 1904–1914. doi: 10.1016/j.chb.2011. 04.015.

Gonzalez, C., & Wong, H. (2012). Understanding stocks and flows through analogy. *System Dynamics Review*, *28*(1), 3-27. doi: 10.1002/sdr.470.

Hardisty, D. J., Johnson, E. J., & Weber, E. U. (2010). A Dirty Word or a Dirty World? Attribute Framing, Political Affiliation, and Query Theory. *Psychological Science, 21*(1), 86-92. doi: 10.1177/0956797609355572.

IPCC. (2007a). Climate Change 2007: Synthesis Report. In R.K. Pachauri & A. Reisinger (Eds.), *Contribution of Working Groups I, II and III to the Fourth Assessment Report of the Intergovernmental Panel on Climate Change* (pp. 104). Geneva, Switzerland: IPCC.

IPCC. (2007b). Climate Change 2007: The Physical Science Basis. In S. Solomon, D. Qin, M. Manning, Z. Chen, M. Marquis, K.B. Averyt, M. Tignor & H.L. Miller (Eds.), *Contribution of Working Group I to the Fourth Assessment Report of the Intergovernmental Panel on Climate Change* (pp. 996), Cambridge, United Kingdom and New York, NY, USA: Cambridge University Press.

Johnson, E. J., & Goldstein, D. (2003). Do defaults save lives? *Science*, 302(5649), 1338-1339.

Johnson, E. J., Hershey, J., Meszaros, J., & Kunreuther, H. (1993).Framing, probability distortions and insurance decisions. *Journal of Risk and Uncertainty*, 7, 35-51.

Jones, A. (2002).*We Can't Afford to "Wait-and-see" on Climate Change.*Retrieved September 7, 2009, from The Sustainability Institute Web site: http://www.sustainer.org/pubs/columns/07.19.02Jones.html.

Leiserowitz, A. (2007). *Public Perception, Opinion and Understanding of Climate Change—Current Patterns, Trends and Limitations.* Retrieved September 7, 2009, from United Nations Development Programme, New York Web site: http://hdr.undp.org/en/reports/global/hdr2007-2008/ papers/leiserowitz_anthony.pdf.

Monbiot, G. (2009). *Climate change: The semantics of denial.* Retrieved November 15, 2009, from The Guardian Web site: http://www.guardian. co.uk/environment/georgemonbiot/2009/feb/27/climate-change-deniers-sceptics.

Moxnes, E., & Saysel, A.K. (2009). Misperceptions of global climate change: information policies. *Climatic Change,93*(1-2), 15-37.

Nordhaus, W. (1994).Expert Opinion on Climatic Change. *American Scientist, 82*, 45-51.

Nordhaus, W. (2008). *A Question of Balance.* New Haven, CT: Yale University Press.

Read, D., Bostrom, A., Morgan, M. G., Fischhoff, B., & Smuts, T. (1994). What do people know about global climate change? 2. Survey studies of educated laypeople. *Risk Analysis*, *14*(6), 959-982.

Singer, S. F. (2009, September 1). Climate Change natural and unstoppable. *FC EDGE,* p. 1.

Sterman, J. D. (1989). Misperceptions of feedback in dynamic decision making. *Organizational Behavior and Human Decision Processes,43*(3), 301-335.

Sterman, J. D. (2000). Learning in and about complex systems.*Reflections: The SoL Journal, 1*(3), 24-51.

Sterman, J. D. (2002). All models are wrong: Reflections on becoming a systems scientist. *System Dynamics Review, 18*, 501-531.

Sterman, J. D. (2008). Risk communication on climate change: Mental models and mass balance. *Science*, *322*, 532-533. DOI: 10.1126/science.1162574.

Sterman, J. D., & Booth Sweeney, L. (2002). Cloudy skies: Assessing public understanding of global warming. *System Dynamics Review, 18*(2), 207-240.

Sterman, J. D., & Booth Sweeney, L. (2007). Understanding public complacency about climate change: Adults' mental models of climate change violate conservation of matter. *Climatic Change, 80*(3-4), 213-238.

Stern, N. (2006). *Stern Review Report on the Economics of Climate Change*. Retrieved September 7, 2009, from HM Treasury Web site: http://www.hm-treasury.gov.uk/sternreview_index.htm.

Taleb, N. N. (2007). *The Black Swan: The Impact of the Highly Improbable*. New York: Random House. ISBN 978-1-4000-6351-2.

Weber, E. U. (2006). Experience-Based and Description-Based Perceptions of Long-Term Risk: Why Global Warming Does Not Scare Us (Yet). *Climatic Change*, *77*(1-2), 103-120.

Reviewed by:

Dr. Cleotilde Gonzalez - Associate Research Professor, Director, Dynamic Decision Making Laboratory (DDMLab), Department of Social and Decision Sciences, Carnegie Mellon University, Pittsburgh, PA, USA. Phone: +1 412-268-6242. Email: coty@cmu.edu.

Dr. Mitchell J. Small - H. John Heinz Professor, Department of Engineering and Public Policy, Department of Civil and Environmental Engineering, Carnegie Mellon University, Pittsburgh, PA, USA. Phone: +1 412-268-8782. Email: ms35@andrew.cmu.edu.

Dr. Lorrie F. Cranor - Associate Professor, Department of Engineering and Public Policy, Department of Computer Science, Carnegie Mellon University, Pittsburgh, PA, USA. Phone: +1 412-268-7534. Email: lorrie@cs.cmu.edu.

Dr.Elke U. Weber - Jerome A. Chazen Professor ofInternational Business, Director, Center for Research on Environmental Decisions, Director, Center for theDecision Sciences, Department of Psychology, Columbia University, New York, NY, USA. Phone: +1 212-854-4427. Email: euw2@columbia.edu.

In: Carbon Dioxide Emissions: New Research
Editors: M. Carpenter and E. J. Shelton
ISBN: 978-1-62257-436-0

Chapter 2

RESPONDING LINEARLY IN NONLINEAR PROBLEMS: APPLICATION TO EARTH'S CLIMATE

Varun Dutt*[*, 1, 2] and Cleotilde Gonzalez[1]
[1] Dynamic Decision Making Laboratory, Department of Social and Decision Sciences, Carnegie Mellon University, Pittsburgh, PA, US
[2] School of Computing and Electrical Engineering, School of Humanities and Social Sciences, Indian Institute of Technology, Mandi, H.P., India

ABSTRACT

Past research has shown that a majority of people exhibit robust linear thinking for nonlinear changes in their decision environment. We argue that linear thinking could be particularly problematic in the case of interpreting carbon-dioxide's (CO_2) lifetime in the earth's atmosphere. Participants from policy and non-policy backgrounds were asked to rank five ranges of CO_2 percentages to be removed from the atmosphere according to their impact on CO_2's lifetime in two separate conditions: Aid and no-Aid. In the Aid condition, participants were provided with a descriptive decision aid through instructions that might enable them to answer the problem correct, while this aid was absent in the no-Aid condition. Two problems were presented to each participant in random order: Linear, where a ranking based upon linear thinking yielded a correct rank order; and Nonlinear, where a ranking based upon linear thinking yielded an incorrect rank order. Results reveal that a majority of participants from both backgrounds responded linearly on both problems and although the decision aid had no effect on participants' correct responses, it enabled policy backgrounds to move away from responding according to linear thinking. We discuss implications of these findings on policymaking about climate change.

[*] Correspondence concerning this article should be addressed to Prof. Varun Dutt, Ph.D., School of Computing and Electrical Engineering, School of Humanities and Social Sciences, Indian Institute of Technology, Mandi, PWD Rest House 2nd Floor, Mandi - 175001, H.P., India. E-mail: varundutt@iitmandi.ac.in.

Keywords: Carbon-dioxide gas's lifetime; linear thinking; climate change; aid; nonlinear problems

INTRODUCTION

According to Galileo Galilei (Galilei, 1638), Aristotle believed that the speed with which an object falls is *linearly* related to its weight. Thus, comparing dropping a ball weighting 100 kg and another weighting 1 kg from the same height, the heavier ball will fall 100 times faster. Responding *linearly* as Aristotle did refers to a function, $f(x) = a*x$, where f(x) is a person's decision response, x is a change in the decision environment, and a is different from zero[1] (Freudenthal, 1983). Many centuries later, Galileo proved Aristotle's reasoning aswrong, but it is unclear whether our tendency to respond linearly to nonlinear problems has been solved.

Currently, there is burgeoning amount of evidence that a majority of people think linearly when encountering nonlinear problems in their decision environment (Cronin, Gonzalez, and Sterman, 2009; Dörner, Kimber, and Kimber, 1997; Dutt and Gonzalez, 2009a; Larrick and Soll, 2008; Van Dooren, De Bock, Janssens, and Verschaffel, 2007). For example, more than 90% of students at the end of elementary schoolresponded "170 seconds" to the question: "John's best time to run 100 meters is 17 seconds. How long will it take him to run 1 kilometer?" (Greer, 1993). Similarly, many people wrongly believe that "3,500 calories consumed is a pound,"or for every 3,500 "extra" calories consumed, you will gain one pound (Chow, 2010).[2] In fact, the tendency to respond linearly has been shown to pervasively affect human judgment in global problems involving serious socio-economic consequences such as those concerning the earth's climate (Dutt and Gonzalez, 2012a; Sterman, 2008; Sterman and Booth Sweeney, 2007). For example, Dutt and Gonzalez (2012b) have shown that when university students were asked to estimate the shape of a carbon-dioxide (CO_2) accumulation and given linear changes in CO_2 emissions and absorption over time, a majority drew a linear shape for the accumulation that was similar to the linear shape of CO_2 emissions overtime. Similarly, Sterman and Booth Sweeney (2007) and Sterman (2008) have shown that people often misperceive the dynamics of CO_2 accumulation; assuming that if one is to increase the accumulation, then CO_2 emissions should increase as well in a shape similar to the accumulation. This tendency to respond linearly is also related to people's level of educationin science and technology (STEM) (Dutt and Gonzalez, 2012b), where people with backgrounds in STEM seem to respond less linearly compared to non-STEM backgrounds.

A prediction that the shape of an accumulation "looks like" the shape of the inflow is an example of robust linear thinking called the *correlation heuristic* (CH) (Cronin et al., 2009). According to Cronin et al. (2009), the proportion of participants relying on the CH increased as the nonlinear relationship between the inflow, outflow, and accumulation became more complex.

In the case of the earth's climate, people may underestimate the extent of the nonlinear increase in CO_2 accumulation (Dutt and Gonzalez, in press). That is because the shape of CO_2

[1] If a = 0, then the relationship is constant (= 0) rather than linear. For a constant relationship, a person's decision response is independent of environmental changes.

[2] The actual relationship between the changes in body weight over time is nonlinear, and is a function of a person's food intake and the difference of one's current body weight from a reference body weight (Chow, 2010).

emissions (inflow) has been increasing about linearly overtime (IPCC, 2007), and people might think that the accumulation will also increase linearly. In practice, an assumption of linear increase will underestimate the actual increase. Furthermore, such underestimations could undermine the urgency of the climate problem and encourage deferment of human actions, leading to *wait-and-see* behavior (Dutt and Gonzalez, 2012a; Sterman, 2008; Sterman and Booth Sweeney, 2007).

It has been argued that overreliance on linear thinking is partly due to its simplicity (Fischbein, 1999; Freudenthal, 1983; Lesh, Post, and Behr, 1988; Rouche, 1989). For example, Rouche (1989) argued that "it is the idea of proportionality that comes immediately in the mind, because undoubtedly there are no functions that are moresimple than the linear ones" (pg. 17). Similarly, Freudenthal (1983) commented that "linearity is such a suggestive property of relations that one readily yields to the seduction to deal with each numerical relation as though it were linear" (pg. 267).

Furthermore, literature on heuristics and biases show that simple linear models lead to approximate correct responses that are more accurate than even expert judgments (Dawes, 1979; Goldberg, 1970). For example, Dawes (1979) gives the example of predicting something as abstract as "professional self-actualization." Given students' graduate record examination, grade point average, and letters of recommendation, one could create a simple linear model to predict the students' professional self-actualization (self-actualization was measured for a set of students based upon their achievementpost-graduation from the university).

When Dawes and Corrigan (1974) applied different linear models to five different datasets to predict the criterion, an equal weighting linear model (the simplest assumption of linearity) out performed all other competing models. Thus, simple linear assumptions can be accurate in many situations, and people depend upon it because it yields an accurate answer in many situations.

Concrete interventions can help reduce linear thinking in both simple and complex nonlinear problems (Cronin et al., 2009; Larrick and Soll, 2008; Garcia-Retamero, Galesic, Gigerenzer, 2010). For example, a physical representation of a nonlinear problem that uses pictures as "metaphors" helped participants reduce their reliance on linear thinking and increased their accuracy (Dutt and Gonzalez, in press). Although these interventions seem to be effective in reducing reliance on linear thinking, they require people to *change* their cognitive thought processes in nonlinear problems, where such a change might at times become very difficult or even impossible to attain (Klayman and Brown, 1993).

This chapter demonstrates robust reliance on linear thinking in a nonlinear environmental problem. It tests concrete interventions to help people respond correctly *without changing* their tendency to think linearly.[3] One intervention is to present a nonlinear problem in a way where linear thinking results in a correct response. Some research has shown that a change in the information context can enable people to make correct responses without influencing their natural thought process (Klayman and Brown, 1993; Payne, Bettman, and Schkade, 1999). Another intervention is to encourage participants through instruction to think nonlinearly; an intervention that tries to change a participant's thought process. Recent research has shown that a *nudge* given in the form of written instructions might enable better decisions (Thaler

[3] We discuss other problems in the discussion section where linear thinking could result in a correct response based upon the problem's presentation.

and Sunstein, 2008). Furthermore, we evaluate whether participants' policy backgrounds influence their reliance on linear thinking in these problems. Because decisions about environmental problems are made by policymakers, it is important to determine if the participant's background in politics, business, economics, and law influence their thinking compared to non-policy backgrounds. According to Nordhaus (1994), the policy background is highly representative of the backgrounds possessed by policymakers who decide on environmental issues facing the world.

The Nonlinear CO_2 Lifetime Problem

The lifetime of CO_2 in the atmosphere (in units of years) is the time it takes to remove a certain mass of CO_2 from the atmosphere. CO_2 lifetime is naturally affected by the yearly percentage of CO_2 removed ("percent-removed" hereafter) by natural processes like absorption by oceans and photosynthesis in plants (IPCC, 2007). A large percent-removed is desirable because larger quantities of accumulated CO_2 leads to climate changeand increasing average temperature (IPCC, 2007). Figure 1 exemplifies CO_2's lifetime in the atmosphere as a nonlinear function of its percent-removed: the lifetime of CO_2 in the atmosphere (units: years) = 100 / percent-removed (units: percent per year).

As shown in Figure 1, a decrease in the percent-removed corresponds to a nonlinear increase in CO_2 lifetime. In addition, the percent-removed is expected to decrease in future years, as oceans and plants are expected to have a reduced ability to absorb CO_2, resulting in a large increase in atmospheric CO_2 lifetime (Cramer et al. 2001; Joos et al. 2001; Matear and Hirst, 1999; Sarmiento and Quéré, 1996).

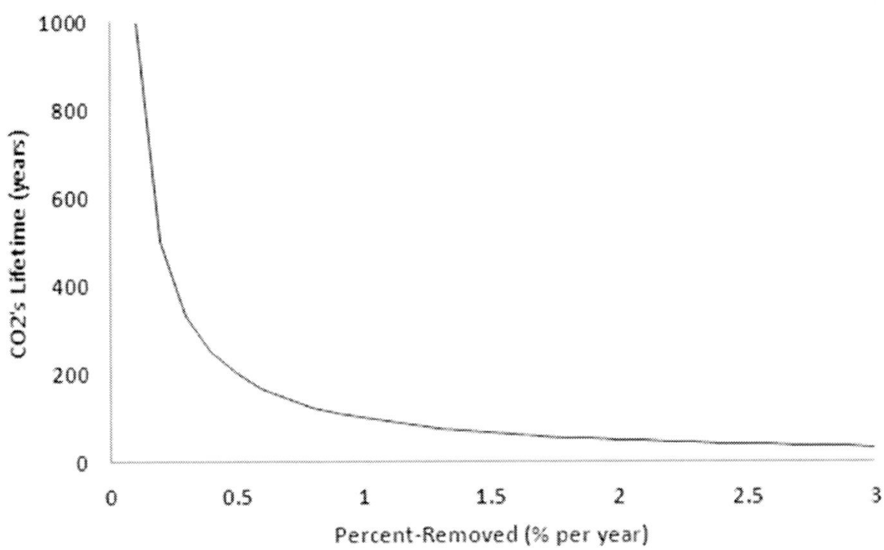

Figure 1. The nonlinear relationship between CO_2's lifetime and percent-removed in the CO_2 lifetime problem.

Given the nonlinear relationship between the percent-removed and CO_2 lifetime, the equal range of reduction in percent-removed may result in a very large or very small increase in CO_2 lifetime, depending on where the range falls on the non-linear curve (see Figure 1). For example, a percent-removed reduction from 0.3 to 0.1 (i.e., 0.2 range) per year results in a 667 years increase for CO_2 lifetime. A reduction from 0.8 to 0.6 (i.e., a similar 0.2 range) per year, however, results in only a 42 years lifetime increase.

Consistent with the substantial evidence of human linear thinking in nonlinear problems, we expect that participants will think linearly when asked to judge the effect of a decrease in CO_2's percent-removed on an increase in CO_2 lifetime: they would believe that the *largest* reduction in percent-removed would result in the *largest* increase in lifetime. Thus, we expect:

- *H1a*: A larger proportion of linear responses than nonlinear responses.

Furthermore, we expect that by presenting a problem where a linear response leads to a correct response, we will enable participants to make correct responses even when relying on linear thinking (Klayman and Brown, 1993; Payne et al., 1999). We accomplish this by changing the presentation of information in the decision environment. This manipulation is strictly in the decision environment, not a treatment to change participants' thought processes. Such anapproach has also been suggested in other judgment research (Larrick, 2004; Klayman and Brown, 1993; Payne et al., 1999). In other cases, however, linear thinking would lead to incorrect responses. Thus, we hypothesize that:

- *H1b*: The proportion of correct responses will be greater when the correct response in the problem is aligned with linear thinking compared to when it is not.

Furthermore, another way to enable participants make better decisions is to provide them with a descriptive aid through instruction (Thaler and Sunstein, 2008). The aid could be in the form of a statement that suggests to convert a CO_2's percent-removed value to a CO_2's lifetime value (where CO_2's lifetime = 100 / percent-removed as seen in Figure 1) and make it simpler for them to calculate the linear increase in CO_2's lifetime. Using CO_2's lifetime information will help a person reduce the nonlinear problem to a linear one, making it easy to answer the problem correctly.

It is to be noted that unlike the above manipulation that changed a person's decision environment, this aid manipulation is aimed at changing a person's linear thought process. Thus, we hypothesize that:

- *H2:* The proportion of correct responses will be greater for those who are given an aid than those who are not given an aid.

Finally, according to Dutt and Gonzalez, (2012b), a greater proportion of STEMs provided correct responses in nonlinear problems compared to non-STEMs. A possible reason for this finding is that STEMs possess greater expertise in mathematical problem solving (Chi, 2006). For example, Chase and Simon (1973) found that expertise and skill in chess enabled participants to recognize significant patterns and remember them easily. Similarly, experience in mathematical problem solving might enable STEMs to respond appropriately in nonlinear problems.

Currently, there is dearth of research that directly investigates people with policy backgrounds' linear responses in nonlinear problems. If policy backgrounds possess some expertise in policymaking, then they should respond more accurately in the CO_2 lifetime problem. Thus, we expect:

- *H3:* A larger proportion of correct responses by those with policy backgrounds compared to those with non-policy backgrounds.

METHODS

Participants

Sixty-seven participants participated in this experiment and were recruited using an online advertisement. Twenty-three participants were from a policy background, and possessed or were pursuing degrees in political science (N=3), business (N=2), economics (N=5), policy (N=9), and law (N=4). The rest of the participants (N=44) had non-policy backgrounds. Thirty-three participants were females. Ages ranged from 18 to 52 years (Mean= 25, SD= 6).

Forty-nine percent of participants were either enrolled in a graduate degree or had completed a graduate degree in the past. Forty-four percent of participants with a policy background and fifty-two percent of participants with a non-policy background were either enrolled in a graduate degree or had completed a graduate degree in the past, respectively. All sixty-seven participants reported knowing some information about climate change through television, radio, newspaper, magazine, movie, or a talk with family or friends. Ten and thirteen participants with policy backgrounds were randomly assigned to the no-Aid and Aid conditions, respectively. Twenty-five and nineteen participants with non-policy backgrounds were randomly assigned to the no-Aid and Aid conditions, respectively. All participants received a flat compensation of $5 in the experiment, which lasted for about 10 minutes.

Materials and Procedure

Each participant was presented with two problems in random order. One of the problems is aligned correctly with linear thinking (Linear) while the other problem is not (Nonlinear). "Aligning correctly" with linear thinking meant that the problem was presented such that a linear response would yield a correct response, while "aligning incorrectly" meant that the problem was presented such that a linear response would yield an incorrect response.

Each problem consisted of five ranges of decreasing values of CO_2 percent-removed per year, with a From (status-quo and higher) and a To (future and lower) value. Participants were asked to rank the percent-removed ranges from the one that would cause the largest increase in CO_2 lifetime (rank 1) to the smallest increase (rank 5) (see Figure 2 for full instructions). Participants were also requested to clearly show their math in the space provided.

Aid Condition, Linear Problem

The table below gives you the pairs of "From" and "To" yearly percentage of CO_2 removed values. Please rank order these values in terms of their detriment to Earth's climate from 1 to 5:

- Use 1 for the **most** detrimental pair: the pair that would cause the most increase of lifetime of CO_2 in the atmosphere.
- Use 5 for the **least** detrimental pair: the pair that would cause the least increase of lifetime of CO_2 in the atmosphere.

You are allowed to use a computer screen calculator, in case you need one. But you also need to clearly show your work (i.e. your mathematical formulation) in the space provided below the table. For calculations, the climate scientist has suggested that you translate the yearly percentage of CO_2 removed values (in percentage of CO_2 per year) into the lifetime that CO_2 stays in the atmosphere (in years).

Pairs		Rank
From	To	(1 to 5)
1.9% per year now	0.5% per year in future	
2.1% per year now	0.1% per year in future	
2.0% per year now	0.3% per year in future	
1.7% per year now	1.1% per year in future	
1.6% per year now	0.9% per year in future	

Please clearly show your work i.e. the math you did for getting the ranks in the table above (you are allowed to use a computer screen calculator for any calculations):

Aid Condition, Nonlinear Problem

The table below gives you the pairs of "From" and "To" yearly percentage of CO_2 removed values. Please rank order these values in terms of their detriment to Earth's climate from 1 to 5:

- Use 1 for the **most** detrimental pair: the pair that would cause the most increase of lifetime of CO_2 in the atmosphere.
- Use 5 for the **least** detrimental pair: the pair that would cause the least increase of lifetime of CO_2 in the atmosphere.

You are allowed to use a computer screen calculator, in case you need one. But you also need to clearly show your work (i.e. your mathematical formulation) in the space provided below the table. For calculations, the climate scientist has suggested that you translate the yearly percentage of CO_2 removed values (in percentage of CO_2 per year) into the lifetime that CO_2 stays in the atmosphere (in years).

Pairs		Rank
From	To	(1 to 5)
0.3% per year now	0.1% per year in future	
0.8% per year now	0.5% per year in future	
2.2% per year now	1.1% per year in future	
0.9% per year now	0.2% per year in future	
2.1% per year now	1.2% per year in future	

Please clearly show your work i.e. the math you did for getting the ranks in the table above (you are allowed to use a computer screen calculator for any calculations):

Figure 2. The climate problems, Linear and Nonlinear, presented to the participants in the Aid condition. The same problems were presented in the no-Aid condition, except that the statement instructing the participant to convert the percent-removed to CO_2's lifetime was omitted.

Participants were randomly assigned to one of two conditions, Aid or no-Aid. In the Aid condition, participants were given the following statement as part of the instructions: "For calculations, the climate scientist has suggested that you translate the yearly percentage of CO_2 removed values (in percentage of CO_2 per year) into the lifetime that CO_2 stays in the atmosphere (in years)." This sentence was omitted from the instructions for participants assigned to the no-Aid condition.

Table 1. Different linear rank orders of the percent-removed ranges in the Linear and Nonlinear problems

Linear Problem									
From	To	Proportional Change	Delta Change	Correct Change in Years	Correct Rule (1)	Difference Rule (2)	Addition Rule (3)	Ratio Rule (4)	Proportional Rule (5)
2.1	0.1	0.95	2.0	952	1	1	1	1	1
2.0	0.3	0.85	1.7	283	2	2	2	2	2
1.9	0.5	0.74	1.4	147	3	3	3	3	3
1.6	0.9	0.44	0.7	49	4	4	4	4	4
1.7	1.1	0.35	0.6	32	5	5	5	5	5
Nonlinear Problem									
From	To	Proportional Change	Delta Change	Correct Change in Years	Correct Rule (1)	Difference Rule (2)	Addition Rule (3)	Ratio Rule (4)	Proportional Rule (5)
2.2	1.1	0.50	1.1	46	4	1	5	3	3
0.9	0.2	0.78	0.7	389	2	3	2	1	1
2.1	1.2	0.43	0.9	36	5	2	4	4	4
0.3	0.1	0.67	0.2	667	1	5	1	2	2
0.8	0.5	0.38	0.3	75	3	4	3	5	5

The ranks and math shown by participants were used to classify the type of procedure they used to respond (linear or nonlinear). Only one sequence of ranks from 1 to 5 is correct response in each problem, however, participants could enter different sequence of ranks by following different rank-order rules. Table 1 provides five different linear rank-order rules that participants could follow in each problem (numbered from 1 to 5) as a result of linear thinking. We made use of these five rules to classify a participant's ranking as being a linear response.

The From and To values are given to participants (in Figure 2). The next three columns: "Proportional Change," "Delta Change," and "Correct Change in Years" are used to calculate five possible linear rules: "Correct Rule," "Difference Rule," "Addition Rule," "Ratio Rule," and "Proportional Rule." Proportional Change refers to the relative change in the percent-removed given by the formula (From - To) / From. Delta Change refers to the difference between the From and To values of a percent-removed range. Correct Change in Years refers to the correct values of CO_2 lifetime that could be obtained by using the formula, 100/To - 100/From. The Correct Rule was the correct rank order obtained through the Correct Change in Years column. The other four rules represent different forms of linear-thinking response: the Difference Rule is the rank order obtained based on the Delta Change column; the Addition Rule is the rank order obtained by the addition of From and To values; the Ratio

Rule is the rank order obtained based on the ratio of From/To; and the Proportional Rule is the rank order obtained using Proportional Change. In the Linear problem, all of the other four rank-order rules are the same as the Correct Rule (or correct response), but not in the Nonlinear problem. Participants' responses were classified according to the rule they appeared to follow, or as "other" if their ranks did not correspond to any of the fivelinear rules (i.e., their responses were nonlinear-incorrect responses). If a participant ranked according to linear response or the Correct Rule in the Linear problem, then this ranking would lead her to a correct response. In contrast, a participant could only get a correct response on the Nonlinear problem by following the Correct Rule. Therefore, following a linear response on the Nonlinear problem could not have produced a correct response.

RESULTS

Two independent raters coded each participant's response as belonging to one the five rank rules (given in Table 1) or as"other". Inter-rater reliability for the two independent raters revealed satisfactory amounts of agreement between the two, Kappa, Correct = 0.94 ($p < 0.001$), 95% CI[1] (0.89, 1.00); Kappa, Difference = 0.97 ($p < 0.001$), 95% CI (0.92, 1.00); Kappa, Addition = 1.00 ($p < .001$), 95% CI (1.00, 1.00); Kappa, Ratio = 0.92 ($p < 0.001$), 95% CI (0.81, 1.00); Kappa, Proportion = 0.92 ($p < 0.001$), 95% CI (0.81, 1.00); and Kappa, Other = 0.93 ($p < 0.001$), 95% CI (0.80, 1.00). These categorizations were used for subsequent analysis of responses after resolving any inconsistency between raters through direct meeting and active discussion.

Proportion of Linear Responses within Each Problem (H1a)

To test H1a, we compared the proportion of linear responses to other (nonlinear) responses within the Linear and Nonlinear problems in the Aid and no-Aid conditions for policy and non-policy backgrounds. Table 2 shows the proportion of correct responses, linear responses, and otherresponses for participants in both problems and both conditions. A non-zero correct response in the Linear problem was only due to linear thinking, and there were 0% correct responses in the Nonlinear problem.

For participants with non-policy backgrounds, the proportion of linear responses was significantly greater than the proportion of other (nonlinear) responses, regardless of the problem or condition: In the Aid condition and Linear problem (84%>16%):$\chi^2(1)=17.789$, $p<.001$, $r=.68$; In the Aid condition and Nonlinear problem (79%>21%): χ^2 (1) =12.737, $p <.001$, $r=.58$; In the no-Aid condition and Linear problem (92%>8%): χ^2 (1)=35.280, $p<.001$, $r=.84$; and in the no-Aid condition and Nonlinear problem (88%>12%): χ^2 (1)=28.880, $p <.001$, $r=.76$.

[1] 95% Confidence interval.

Table 2. Proportion of participants following a correct, linear, and other response in the experiment

	Policy Backgrounds				Non-policy Backgrounds			
	Aid		No-Aid		Aid		No-Aid	
Response	Linear (%)	Nonlinear (%)	Linear (%)	Nonlinear (%)	Linear (%)	Nonlinear (%)	Linear (%)	Nonlinear (%)
Correct	62	00	50	00	63	00	72	00
Linear	62	54	80	80	84	79	92	88
Other	38	46	20	20	16	21	08	12

These results supports hypothesis H1a. For participants with policy backgrounds, the proportion of linear responses was significantly greater than the proportion of other (nonlinear) responses in the no-Aid condition's Linear (80%>20%) (χ^2 (1)=7.200, p<.01, r=.60) and Nonlinear problem (80%>20%)(χ^2 (1)=7.200, p<.01, r=.60). However, there was no difference between the proportion of linear responses and other responsesin the Aid condition's Linear and Nonlinear problem (Linear problem: linear response (62%) = other response (38%) with χ^2 (1)=1.385, *ns*, r =.23; Non-linear problem: linear response (54%)=other response (46%) with χ^2 (1)=0.154, *ns*, r=.08).

Therefore, an aid helped participants with policybackgrounds to rely less on linear responses. Support for hypothesis H1a is present in the problem without Aid, but not in the problem with Aid.

Proportion of Correct and Linear Responses between Linear and Nonlinear Problems (H1b)

To test H1b, we compared the proportion of correct responses between each problem in each condition for policy and non-policy backgrounds, respectively. For non-policy backgrounds, the proportion of correct responses was significantly greater for the Linear problem compared to the Nonlinear problem in both conditions (see Table 2) (Aid: 63% > 0% with χ^2 (1)=17.538, p< .001, r = .68; no-Aid: 72% > 0% with χ^2 (1) = 28.125, p< .001, r=.75).

Similarly, for policy backgrounds, the proportion of correct responses was significantly greater for the Linear problem compared to the Nonlinear problem in both conditions (Aid: 62% > 0% with χ^2 (1) = 13.765, p< .001, r=.73; no-Aid: 50% > 0% with χ^2 (1) = 6.667, p< .01, r = .58). These results support H1b.

Furthermore, regardless of the background, the proportion of participants giving linear responses was no different between each problem in both conditions (see Table 2) (For non-policy background: Aid: 84% = 79% with χ^2 (1) = 0.175, *ns*, r = .07; no-Aid: 92% = 88% with χ^2 (1) = 0.222, *ns*, r=.07. For policy background: Aid: 62% = 54% with χ^2 (1) = 0.158, *ns*, r=.08; no-Aid: 80% = 80% with χ^2 (1) = 0.000, *ns*, r = .00). These results show that the difference in correct responses between problems was due to the participants' persistent reliance on linear reasoning, regardless of their backgrounds and any aid.

Proportion of Correct and Linear Responses between Aid and No-Aid Conditions (H2)

To test H2, we compared the proportion of correct responses between the Aid and no-Aid conditions in the Linear and Nonlinear problems for non-policy and policy backgrounds. Aid had no effect on the proportion of correct responses in the Linear problem (for non-policy background: Aid: 63% = no-Aid: 72% with χ^2 (1) = 0.389, *ns*, r = .09; for policy background: Aid: 62% = no-Aid: 50% with χ^2 (1) = 0.878, *ns*, r = .20) or in the Nonlinear problem (for non-policy background: Aid: 0% = no-Aid: 0% with χ^2 (1) = no-statistic[2], *ns*, r = no-statistic; for policy background: Aid: 0% = no-Aid: 0% with χ^2 (1) = no-statistic, *ns*, r = no-statistic). Again, Aid had no effect on the proportion of linear responses in the Linear problem (for non-policy background: Aid: 84% = no-Aid: 92% with χ^2 (1) = 0.650, *ns*, r = .12; for policy background: Aid: 62% = no-Aid: 80% with χ^2 (1) = 0.910, *ns*, r = .20) or in the Nonlinear problem (for non-policy background: Aid: 79% = no-Aid: 88% with χ^2 (1) = 0.661, *ns*, r=.12; for policy background: Aid: 54% = no-Aid: 80% with χ^2 (1) = 1.704, *ns*, r = .27). Thus, Aid had no influence on participants'reliance on linear or correct responses. These results do not support hypothesis H2.

Effects of Educational Backgrounds

Finally, to test hypothesis H3, we compared the proportion of correct responses between policy and non-policy backgrounds in each problem in the Aid and no-Aid conditions, respectively. Overall, there was no difference for participants with policy and non-policy backgrounds. This finding holds in the Aid condition for the Linear problem (Correct response: 62% = 63% with χ^2 (1) = 0.126, *ns*, r = .06; Linear response:62% = 84% with χ^2 (1) = 2.116, *ns*, r = .26) and for the Nonlinear problem(Correct response: 0% = 0%with χ^2 (1) = no-statistic, *ns*, r = no-statistic; Linear response: 54% = 79% with χ^2 (1) = 2.264, *ns*, r = 0.27). This finding also holds in the no-Aid condition for the Linear problem (Correct response: 50% = 72% with χ^2 (1) = 1.534, *ns*, r = .21; Linear response: 80% = 92% with χ^2 (1) = 1.016, *ns*, r = .17) and for the Nonlinear problem (Correct response: 0% = 0% with χ^2 (1) = no-statistic, *ns*, r= no-statistic; Linear response: 80% = 88% with χ^2 (1) = 0.373, *ns*, r=.10). When taken together, these results do not support hypothesis H3.

General Discussion

This research shows that people's linear thinking is pervasive while making judgments in nonlinear environmental problems. Our manipulation of aligning correct responses with linear thinking proved highly effective. Changing the information in the problem to align with a person's dominant decision-making strategy (which in this case is responding linearly) can be an effective way of improving their decision making (Klayman and Brown, 1993; Payne et

[2] Because there is no participant in the Nonlinear problem who gave a correct response, there is no statistic to report for the comparison due to the absence of data.

al., 1999). Moreover, the important point to note is that the alignment manipulation does not change their thought processes in any way. The manipulation is simply meant to make use of these linear cognitive processes to help participants understand nonlinear problems in the way they naturally are inclined to and thus enable them to correctly respond. Furthermore, our results agree with prior evidence of linear thinking in environmental problems concerning inferences about CO_2 accumulation (Sterman, 2008; Sterman and Booth Sweeney, 2007).

Our results also indicate that a majority of participants with non-policy backgrounds responded linearly in problems with or without instructional aid and regardless of whether or not correct response was aligned with linear thinking. There could be a number of reasons for this over reliance amongst participants with non-policy backgrounds. First, it could simply be because linear responses are the simplest response to come to mind (Fischbein, 1999; Freudenthal, 1983; Lesh, Post, and Behr, 1988; Rouche, 1989). Second, literature has shown that even simple linear models lead to correct approximate responses in many cases that are more accurate than expert judgments (Dawes, 1979; Goldberg, 1970). Linear thinking offers two crucial benefits of being simple and/orit producing accurate and good enough answers in many problems, while avoiding more complicated nonlinear rules.

Furthermore, in our results, participants with policy backgrounds still relied on linear thinking and were not able to provide correct responses. But they were able to make some use of the instructional aid, which helped them to move away from linear thinking to other nonlinear or incorrect types of response. Although we can only currently speculate, one reason may be that it is challenging to *change* participants' cognitive thought processes to enable better decisions (Klayman and Brown, 1993) and the aid was not inadequate in producing this change.

A second reason could be on account of the aid's effectiveness itself: a descriptive aid that provided participants with the exact relationship between CO_2 lifetime and percent-removed might have been more effective. Still, participants with policy backgrounds, just like those with non-policy backgrounds, are limited by their cognitive capacity (Sterman and Booth Sweeney, 2007) and are thus unable to utilize the aid effectively. We plan to investigate these explanations as part of future research.

It is expected that the yearly percentage of CO_2 removed from the atmosphere will decrease in future years, resulting in a large increase in CO_2 lifetime (Cramer et al., 2001; Joos et al. 2001; Matear and Hirst, 1999; Sarmiento and Quéré, 1996). As this change will be detrimental to the earth's climate, accurate human assessment of the nonlinear relationship between the percent-removed and CO_2 lifetime is important. When participants were given a problem where a linear response would lead to an incorrect answer, none of the participants in the experiment correctly ranked the decreasing percent-removed ranges in the problem. They ultimately underestimate the most detrimental changes in CO_2 lifetime. This inaccurate assessment could be a possible reason for wait-and-see policies for climate change.

Finally, aligning a nonlinear problem with linear mental models is a manipulation that may also be useful in many other important problems. For example, an intervention similar to the one tested here could be attempted for reducing the dispersal of a commodity (e.g., pollution in river) by giving people choices about the taxes they pay per unit of dispersing the commodity. For example, consider a certain tax per kilogram on pollution created in a river (units: $/Kg of pollution) aimed at reducing pollution. Polluters, like large industrial factories on the river's banks, could be offered different taxation choices, where they are charged with a smaller tax now and a larger tax in the near future for each policy. A range of tax increases could be designed in such a way that the smallest increase for the same amount of total pollution appears the most attractive to polluters according to their linear thinking. But in fact, the smallest increase has the maximum potential to reduce river pollution. Our future endeavor in this research will be to extend the problem alignment intervention to other nonlinear problems faced in daily life.

ACKNOWLEDGMENTS

This research is supported by the Defense Threat Reduction Agency (DTRA) grant number: HDTRA1-09-1-0053 to Dr. Cleotilde Gonzalez. The authors would like to thank Dr. Lorrie F. Cranor, Dr. Jolie M. Martin, and Dr. Matthew A. Cronin for their help in reviewing this manuscript and providing insightful comments.The authors are also thankful to Ms. Hau-yu Wong, DDMLab, Carnegie Mellon University for helping in coding participant explanations and providing editorial comments.

REFERENCES

Chase, W.G., & Simon, H.A. (1973). Perception in chess. *Cognitive Psychology, 4*(1), 55-81. DOI: 10.1016/0010-0285(73)90004-2.

Chi, M.T.H. (2006). Methods to assess the representations of experts' and novices' knowledge. In K.A. Ericsson, N. Charness, P. Feltovich, and R. Hoffman (Eds.), *Cambridge Handbook of Expertise and Expert Performance* (pp. 167-184).Cambridge, MA: Cambridge University Press.

Chow, C. (2010). Obesity, weight gain and a cookie. Retrieved from the Scientific Clearing.

House website: http://sciencehouse.wordpress.com/2010/03/04/obesity-weight-gain-and-a-cookie/.

Cramer, W., Bondeau, A., Woodward, F. I., Prentice, C., Betts, R. A., Brovkin, V., et al. (2001).

Global response of terrestrial ecosystem structure and function to CO_2 and climate change: Results from six dynamic global vegetation models. *Global Change Biology,7*(4), 357-373. DOI: 10.1046/j.1365-2486.2001. 00383.x.

Cronin, M., Gonzalez, C., & Sterman, J. D. (2009). Why don't well-educated adults understand accumulation? A challenge to researchers, educators and citizens. *Organizational Behavior and Human Decision Processes, 108*(1), 116-130. DOI:10. 1016/j.obhdp.2008.03.003.

Dawes, R. M. (1979). The robust beauty of improper linear models in decision making. *American Psychologist, 34*(7), 571-582. DOI:10.1037/0003-066X.34.7.571.

Dawes, R. M. & Corrigan, B. (1974). Linear models in decision making. *Psychological Bulletin, 81*(2), 95-106. DOI: 10.1037/h0037613.

Dörner, D., Kimber, R., Kimber, R. (1997). *The logic of failure: Recognizing and avoiding error in complex situations.*Cambridge, MA: Basic Books.

Dutt, V., & Gonzalez, C. (2012a). Human Control of Climate. *Climatic Change, 111*(3), 497-518. DOI: 10.1007/s10584-011-0202-x.

Dutt, V., & Gonzalez, C. (2012b). Decisions from Experience reduce misconceptions about Climate Change. *Journal of Environmental Psychology*, *32*(1), 19-29.

Dutt, V., & Gonzalez, C. (in press). Reducing the Linear Perception of Non-Linearity: Use of a Physical Representation. *Journal of Behavioral Decision Making*. DOI: 10.1002/bdm.759.

Fischbein, E. (1999). Intuitions and schemata in mathematical reasoning. *Educational Studies in Mathematics, 38*(1-3), 11–50. DOI: 10.1023/A: 1003488222875.

Freudenthal, H. (1983). *Didactical phenomenology of mathematical structures*. Dordrecht, the Netherlands: Reidel.

Galilei, G. (1638). *Discorsi e dimostrazioni matematiche intorna a due nuove scienze: 1954, Dialogues concerning two new sciences*. New York: Dover.

Garcia-Retamero, R., Galesic, M., & Gigerenzer, G. (2010). Do icon arrays help reduce denominator neglect? *Medical Decision Making*, *30*(6), 672-684. DOI: 10.1177/0272989X10369000.

Goldberg, L. R. (1970). Man versus model of man: a rationale, plus some evidence, for a method of improving on clinical inferences. *Psychological Bulletin*, *73*(6), 422-432. DOI: 10.1037/h0029230.

Greer, B. (1993). The mathematical modelling perspective on wor(l)d problems. *Journal of Mathematical Behavior, 12*(3), 239–250.

IPCC. (2007). Climate change 2007: Synthesis report. In R. K. Pachauri, & A. Reisinger (Eds.), *Contribution of working groups I, II and III to the fourth assessment report of the intergovernmental panel on climate change* (pp. 1–104). Geneva, Switzerland: IPCC.

Joos, F., Prentice, I. C., Sitch, S., Meyer, R., Hooss, G., Plattner, G. K., Gerber, S., & Hasselmann, K. (2001). Global warming feedbacks on terrestrial carbon uptake under the Intergovernmental Panel on Climate Change (IPCC) emission scenarios. *Global Biogeochemical Cycles*, *15*(4), 891-907. DOI:10.1029/2000GB001375.

Klayman, J., & Brown, K. (2002). Debias the environment instead of the judge: An alternative approach to reducing error in diagnostic (and other) judgment. *Cognition*, *49* (1-2), 97-122. DOI:10.1016/0010-0277(93) 90037-V.

Larrick, R. P. (2004). Debiasing. In D. J. Koehler & N. Harvey (Eds.), *Blackwell Handbook of Judgment and Decision Making* (pp. 316-337), Malden, MA: Blackwell.

Larrick, R. P., & Soll, J. B. (2008). The MPG illusion. *Science*, *320*(5883), 1593-1594. DOI:10.1126/science.1154983.

Lesh, R., Post, T., & Behr, M. (1988). Proportional reasoning. In J. Hiebert & M. Behr (Eds.), *Number concepts and operations in the middle grades* (pp. 93–113). Reston, VA: National Council ofTeachers of Mathematics.

Matear, R. J. & Hirst, A. C. (1999). Climate change feedback on the future oceanic CO_2 uptake. *Tellus, 51*(3), 722-733. DOI: 10.1034/j.1600-0889.1999.t01-1-00012.x.

Nordhaus, W. (1994). Expert opinion on climatic change. *American Scientist,82*, 45-51.

Payne, J.W., Bettman, J.R., & Schkade, D.A. (1999). Measuring constructed preferences: Towards a building code. *Journal of Risk and Uncertainty*, *19*(1-3), 243-270. DOI: 10.1023/A:1007843931054.

Rouche, N. (1989). Prouver: Amener à l'évidence ou contrôler des implications? [Proving: Bringing back to evidence or verifying the implications?]. In Commission inter-IREM Histoire et Epistémologie des Mathématiques (Ed.), *La démonstration dans l'histoire* (pp. 8–38). Lyon, France: Institut de Recherche sur l'Enseignement des Mathématiques.

Sarmiento, J. L., & Quéré, C. L. (1996). Oceanic carbon dioxide uptake in a model of century-scale global warming. *Science*, *274*(5291), 1346-1350. DOI: 10.1126/science.274. 5291.1346.

Sterman, J. D. (2008). Risk communication on climate: Mental models and mass balance. *Science, 322* (5901), 532-533. DOI: 10.1126/science. 1162574.

Sterman, J. D., & Booth Sweeney, L. (2007). Understanding public complacency about climate change: Adults' mental models of climate change violate conservation of matter. *Climatic Change, 80*(3-4), 213-238. DOI: 10.1007/s10584-006-9107-5.

Thaler, R. H., & Sunstein, C. R. (2008). *Nudge: Improving decisions about health, wealth, and happiness*. New Haven, CT: Yale University Press.

Van Dooren, W., De Bock, D., Janssens, D., & Verschaffel, L. (2007). Pupils' over-reliance on linearity: A scholastic effect? *British Journal of Educational Psychology*, *77*(2), 307-321. DOI: 10.1348/000709906 X115967.

Reviewed by:

Dr. Lorrie F. Cranor - Associate Professor of Computer Science and Engineering and Public Policy, Carnegie Mellon University, CIC 2207, 4720 Forbes Avenue, Pittsburgh, PA 15213. Phone: +1 412-268-7534. Email: lorrie@cs.cmu.edu.

Dr. Jolie M Martin - Assistant Professor of Journalism and Mass Communication, University of Minnesota, 111 MurH, 206 Church Street, Minneapolis, MN 55455. Phone: +1 617-851-6270. Email: jolie@umn.edu.

Dr. Matthew A. Cronin - Associate Professor of Management, George Mason University, Enterprise Hall 216, 4400 University Dr. Fairfax, VA 22030. Phone: +1 703-993-1783. Email: mcronin@gmu.edu.

In: Carbon Dioxide Emissions: New Research
Editors: M. Carpenter and E. J. Shelton
ISBN: 978-1-62257-436-0

Chapter 3

STEEL INDUSTRY AND CARBON DIOXIDE EMISSIONS: A NOVEL IRONMAKING PROCESS WITH GREATLY REDUCED CARBON FOOTPRINT

H. Y. Sohn and Moo Eob Choi
Department of Metallurgical Engineering, University of Utah, Salt Lake City, UT, US

ABSTRACT

The steel industry is faced with a wide range of environmental concerns that are related to CO_2 emissions. Increasing demand to minimize the greenhouse gas emissions has intensified the pressure on steelmakers and significantly impacted the direction of the development of this industrial sector.

The mitigation of carbon dioxide emissions in the steel industry has been preformed by modifying and integrating currently available technolgies, especially more efficiently utilizing the heavily fossil-oriented energy sources. Although CO_2 emissions per ton steel produced has been noticeably reduced during the past decade, the steel industry still accounts for a significant portion of anthropogenic emissions of carbon dioxide and the modern steel plants operate close to the limits of what is technically possible.

Further reduction of carbon dioxide emissions is not possible without drastic technological changes. Here, the efforts of the steel industry on the development of new transformational technologies will be presented, especially introducing a novel flash ironmaking technology being developed at the University of Utah under the American Iron and Steel Institute's CO_2 Breakthrough Program. This technology will greatly reduce carbon dioxide emissions in the ironmaking step by eliminating coke oven and the pelletization or sintering step and by the use of hydrogen or natural gas as reductant/fuel. When it is operated with natural gas, the new technology will generate only 39% of carbon dioxide produced from the average blast furnace process.

INTRODUCTION

The steel industry is faced with a wide range of environmental concerns that are fundamentally related to CO_2 emissions, high energy requirement, material usage, and byproducts associated with large amount of steel production worldwide. The impact of environmental regulations on steel companies is largely in the ironmaking step because more than 80% of energy is consumed and most pollutants are created in these operations (Lu, 1997).

As climate changes, reportedly accelerated by man-made carbon emissions, have become a significant global issue, increasing demand to minimize greenhouse gas (GHG) emissions has intensified the pressure on steel makers and certainly impacted the direction of the development of the steel industry. The mitigation of CO_2 emission, having the highest global warming effects and accounting for half the effects of all GHGs (Chang-qing et al., 2006; Karl et al., 2010) in the steel industry, has been performed by modifying and integrating currently available technologies, especially more efficiently utilizing the heavily fossil-oriented energy sources. Examples include a less energy-consuming bonding technology for pelletization, prolonging of coke-oven life and reduction of pollutant emissions in coke-making, and increasing of the campaign life of blast furnace (BF), and the pulverized coal injection (PCI) rate (Emi and Seetharaman, 2000).

Although CO_2 emissions per ton steel has been noticeably reduced thanks to such efforts during the past decades, the steel industry still contributes around 5-7 % to the total anthropogenic emissions of CO_2 (Kim and Worrell, 2002; Gielen, 2003; Tacke and Steffen, 2004). Since modern steel plants operate very close to the limits of what is technically possible, it is obvious that further reduction of CO_2 emission is not possible without drastic technological changes. It is envisioned that an entirely new transformational Flash Ironmaking Process described in this chapter would be a viable alternative process for producing iron that addresses the environmental concerns and limitations associated with the blast furnace technology and other available direct reduced iron processes.

Technology Description of Flash Ironmaking

The Flash Ironmaking Technology is based on the direct gaseous reduction of iron oxide concentrates. The technology has the potential to eliminate cokemaking and sintering/ pelletization in the predominant ironmaking technology (BF furnace) thereby lowering carbon dioxide emissions by 61-96% and reducing energy consumption by 32-57%, depending on the choice of fuel/reductant between natural gas and hydrogen, vs. the average blast furnace ironmaking route.

An important factor in the development of the Flash Ironmaking Process is that the U.S. already produces large quantities of iron oxide concentrates that are well suited for flash reduction. These concentrates have to undergo pelletization or sintering process to be utilized as a feed to the blast furnace. Up to 60% of the U.S. iron production (USGS, 2005) is based on taconite concentrates (50-55 million tons/year). On a world-wide basis, good quality lump ore is at a premium. Ores from newer reserves also need to be ground to finer sizes in order to beneficiate them and make them acceptable as feed materials for ironmaking.

The Flash Ironmaking Process is to be operated using natural gas as the main fuel and reducing agent. Natural gas is a readily available resource in the U.S. According to the 2011 Annual Energy Outlook (USEIA, 2011a), the estimations on the U.S. production rate of natural gas predict a steady increase. Further, there is a strong push to increase its gas production in the U.S. from large reserves. The recently discovered reserves in Marcellus Shale (Schwartzel, 2011; USEIA, 2011b) represent a good example of this trend. For an annual production of 20 million tons of iron, which would represent about 40% of the current U.S. production rate, the Flash Ironmaking Process would require 0.33 TCF/year (trillion cubic feet per year) of natural gas. This represents less than 1.5% of the total U.S. consumption rate (22.7 TCF/year) (USEIA, 2011a). When natural gas is used as a reducing agent, the proposed process will produce iron with varying carbon contents such as the iron produced by HYL's ZR self-reforming process (Duarte et al., 2010). In the latter process, carbon levels can be up to 5.5 %. This provides significant advantages in the EAF, such as a lower heat requirement and ease of foamy slag generation.

The Flash Ironmaking Process can also be designed to be operated with pure hydrogen when the anticipated development of hydrogen economy is realized. The product will be low-carbon iron with a possibility for direct steelmaking in the same unit, as shown in Figure 1. The product could also be collected as a solid to be fed into an EAF for steelmaking, in which case carbon could be added in the briquetting step. Alternatively, the EAF could be modified by adding a lance for gas injection to promote optimum slag foaming, rather than relying on CO produced from dissolved carbon, to enhance the refining and steelmaking reactions. Further, the reduced iron material could be melted in a coal-based melter or an electric melter to produce a molten metal feed of different characteristics.

The Flash Ironmaking Process will remove many of the limitations associated with other alternative processes. Specifically, (a) direct use of iron oxide concentrates without the need for pelletization or sintering; (b) no cokemaking required (If coal is used to generate the hot reducing gas, pulverized coal of wide variety can be used.); (c) high temperature can be used because there will be no particle sticking or fusion problems (and thus, the process can be intensive); (d) possibility to produce either solid or molten iron; and (e) low refractory problems, ease of feeding the raw materials, and the possibility of direct steelmaking in a single unit, as shown in Figure 1.

Based on the above discussion, a flash reduction process presents an excellent alternative for a high-intensity process. The fineness of the concentrate particles allows a very rapid reaction rate, thus requiring residence time measured in seconds compared with the minutes and hours it takes to reduce pellets and even fines.

Historically, processes based on iron ore *fines* have been less than successful. One must recognize, however, that *concentrates* are much finer, by more than one or two orders of magnitude, than the typical iron ore 'fines' used in a typical fluidized bed, and the proposed process is meant to take advantage of this fact. The fluid bed processes use particles in the range of +0.1 mm to -10 mm (Iron Carbide, Circored and FIOR) compared with 0.02 – 0.06 mm particles in the concentrates. The rate of reduction increases in proportion to the reciprocal of the particle size or its square, depending on temperature. Thus, compared with concentrate particles, even ore 'fines' take much longer time to reduce than the time available in a flash reduction process. Experimental data obtained from the kinetics feasibility tests have conclusively proven that concentrate particles are sufficiently small to permit the use of a flash reduction process with its short residence time (Sohn et al., 2009; Wang, 2011). In that work, it was shown that more than 95% reduction of iron oxide concentrate particles could be obtained by hydrogen or a mixture of carbon monoxide and hydrogen in 2-7 seconds at temperatures of 1200°C or above.

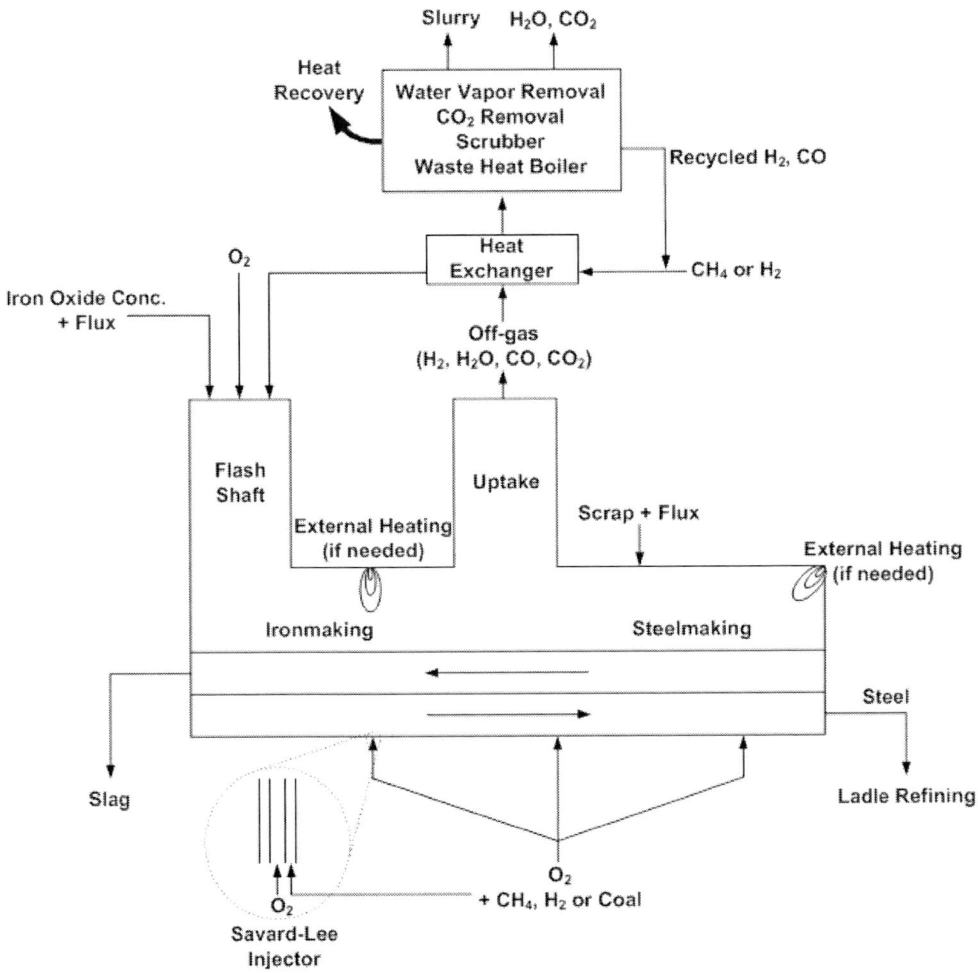

Figure 1. A schematic diagram of a possible direct steelmaking process based on Flash Ironmaking Process.

High temperature gas-solid suspension contact processes, such as the flash, flame reaction or cyclone smelting, are well established in industry and can readily provide a residence time of up to about 10 seconds. The success enjoyed by the flash smelting processes in nonferrous industries, which use concentrate particles of similar sizes, further supports the expectation of the success of the proposed technology. Therefore, it is necessary to develop and test a technically sound process concept that takes advantage of the rapid reduction of iron concentrates at a large scale.

Figure 2 shows a schematic of the evolution of the project objectives and facilities used or to be used in various project phases, leading to commercialization. Currently, a large bench reactor is under development at the University of Utah.

Project Phases	Kinetic Feasibility	Tests with Utah Flash Reactor & Design of Bench Reactor	Test with Bench Reactor	Industrial-Scale Pilot Plant Tests
Experimental Apparatuses	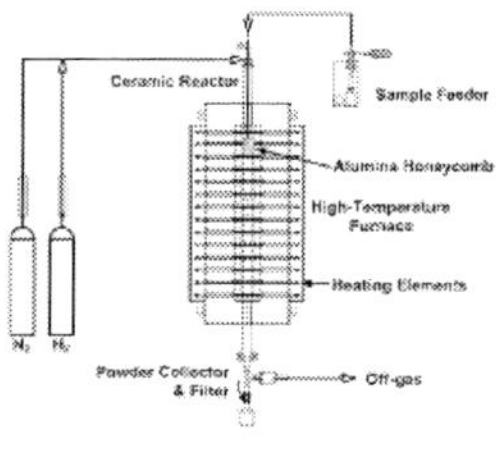	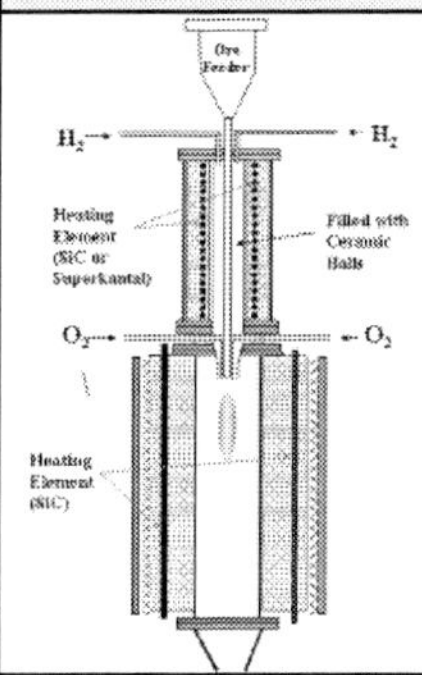	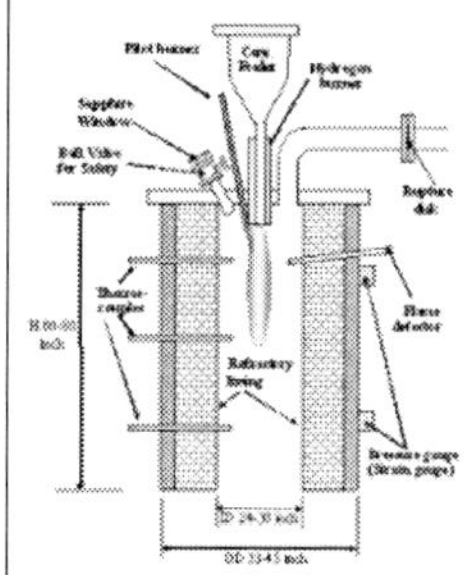	1. Proceed directly to: 75-100k tpy 2. Proceed to modest-scale: 10-25k tpy 3. Expand U of Utah work: Similar to bench reactor but larger
Description	**Drop-Tube Reactor (DTR):** Gas is fed uniformly across the entire cross-sectional area.	**Utah Flash Reactor:** A steel tube inside a thick refractory tube heated with a H_2/O_2 burner and externally by electricity. Limited solid feed rate and temperature.	**Bench Reactor:** A refractory-lined steel vessel. Heat is supplied by a burner and a plasma gun. Concentrate feed rate ~5 kg/h.	**Industrial-Pilot Scale Reactor:** Selected approach would depend on funding availability.

Figure 2. Overall Technical Approach of the Flash Ironmaking Process.

CO_2 MITIGATION AND ENERGY BENEFIT

Preliminary material and energy balance calculations for the Flash Ironmaking Process were performed and presented in a previous publication (Sohn et al., 2009). They were conducted for three possible fuel types, i.e. hydrogen (H_2), natural gas (CH_4), and bituminous coal (considered as $C_{1.4}H$), used with the new technology, together with those for the average blast furnace operation. The generation of CO_2 for the new technology was greatly reduced to ~4% of the average blast furnace with hydrogen as fuel/reductant and moderately reduced to ~70% with the use of coal. It was also shown that using any of the three possible reductants and fuels requires a much smaller amount of energy with ~40% reduction compared with the average blast furnace, which was mainly attained from the elimination of cokemaking/ burning and the pelletization and sintering steps.

Encouraged by these results, more detailed process design and simulation of a commercial-scale process with an annual production rate of 1 million tons based on natural gas or hydrogen has been performed using METSIM, metallurgical process simulation software. Energy requirement and carbon dioxide emissions of each process were compared with the average blast furnace process. It is noted that more detailed information, such as the definitions of all the basic terms, the procedure for the flow sheet construction, the constructed block diagrams for each process route, the reactions considered in each unit process, and the economic feasibility analysis using net present value (NPV) performed based on the results of the process simulation, can be found in elsewhere (Pinegar et al., 2011; 2012).

CO_2 and Other Pollutant Emissions

The Flash Ironmaking Process will have significant advantages over the current blast furnace technology in terms of greenhouse gas and other pollutant emissions. Based on detailed process simulations, the material balance for a commercial-scale Flash Ironmaking Process was compared with the average current blast furnace process (Pinegar et al., 2011; 2012).

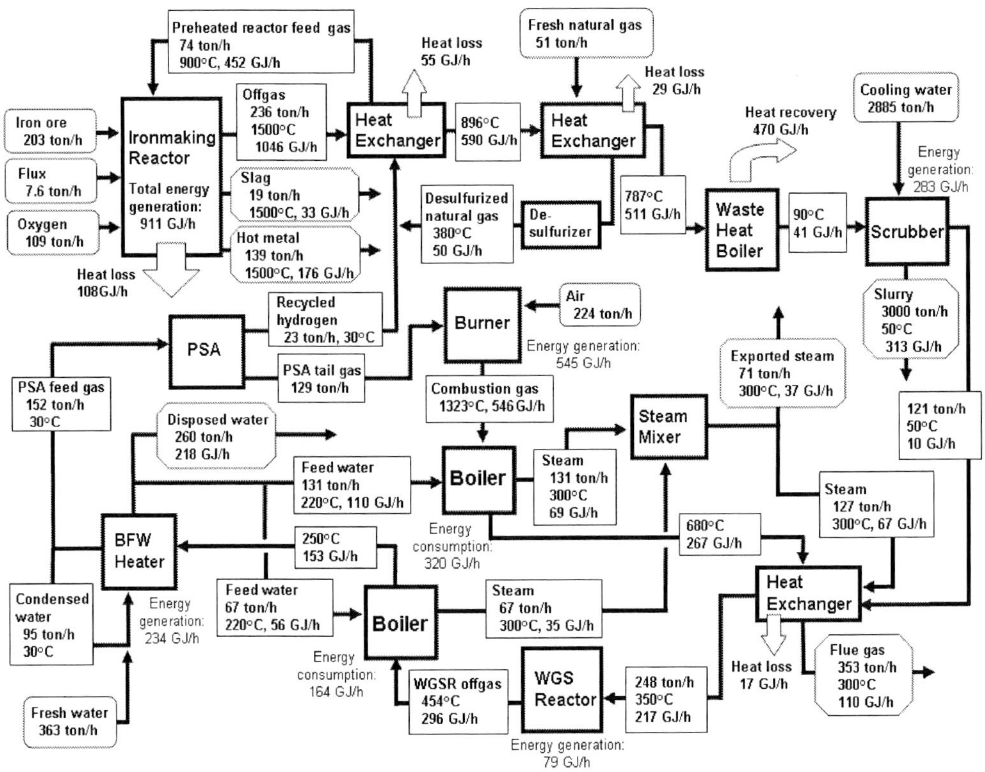

Figure 3. Material and energy flows for the commercial-scale Flash Ironmaking Process producing 1 million metric tons of iron annually with the operating temperature of 1500°C (300 days operation in 1 year). (All 'tons' are metric tons in this report.) [Adapted from Pinegar et al. (2012 PartI)]

Figure 3 shows material and energy flows for a commercial-scale Flash Ironmaking Process based on the direct charging of natural gas (without external reforming). Such a process in which natural gas is directly fed into an ironmaking reactor without reforming is currently practiced by Tenova HYL. A summarized comparison of material balance for the Flash Ironmaking Process using natural gas and the average blast furnace process is presented in Table 1.

For the case of operating the Flash Ironmaking Process with hydrogen instead of natural gas, Table 2 shows a comparison of the hydrogen gas requirement and carbon dioxide emissions with the average blast furnace process.

Table 1. Material balance of a commercial-scale Flash Ironmaking Process using natural gas compared with the average blast furnace process

		Flash Process using natural gas	BF process
INPUT (tons/ton of iron)	Natural gas	0.37	
	Oxygen	0.78	0.68
	Iron ore	1.46 (magnetite)	1.43 (hematite)
	Flux	0.06	0.29
	Nitrogen		2.21
	Coke		0.46
	Fresh water	2.61	
	Air	1.61	
	Total	6.89	5.07
	Cooling water	20.8	22.3 **
OUTPUT (tons/ton of iron)	Iron	1.00	1.05 (contains 4.5% C)
	Slag	0.14	0.21
	Condensed water vapor in slurry	0.83	
	Hot water not used	1.87	
	Steam not used	0.51	
	Total discharged gas	2.54	3.81
	Carbon dioxide	0.98	1.60 *
	Nitrogen	1.25	2.21
	Total	6.89	5.07
	Cooling water in slurry	20.8	22.3 **
	Carbon dioxide from $CaCO_3$ and $MgCO_3$ calcination	0.04	

* This number does not include the carbon dioxide emissions from ore/coke preparation.

** Water requirement for US Steel's Fairfield No. 8 Blast Furnace (production rate: 5500 tons/day) (U.S. Steel Corp., 1985; Camlic et al., 1997). As the reference years of water usage is older than the year of the production rate, this number may be somewhat larger than the number for a modern plant.

Table 2. A comparison of hydrogen gas requirement and carbon dioxide emissions between the H_2-based Flash Ironmaking Process and the average blast furnace process

	H_2-based process	BF process
Hydrogen gas requirement (tons/ton of iron)	0.10	
Carbon dioxide emissions (tons/ton of iron)	0.04 [a]	1.60 [b]

[a] Carbon dioxide emissions from flux calcination are included.

[b] Carbon dioxide emissions from ore/coke preparation for BF process are (~ 0.1 tons/ton of iron) not included.

The results in Tables 1 and 2 indicate that the Flash Ironmaking Process would have significant advantages over the current blast furnace technology as follows:

- When natural gas is used, it will emit 39% less carbon dioxide.
- The use of hydrogen as reductant/fuel emits almost no carbon dioxide or other pollutants in the ironmaking step.

With the elimination of the cokemaking and pelletizing/sintering steps, the associated environmental impact will also be removed in the Flash Ironmaking Process. Typical amounts of pollutant emissions from the average current blast furnace operation are shown in Table 3. As noted in this table, the amounts from cokemaking do not include the possibly large amounts of pollutants from various other steps for which no data are available, according to the report cited. A particularly serious absence is the organic pollutants dissolved in the quench water used in cokemaking.

The requirement of fuel supply to the blast furnace in the form of oil, natural gas, and coke breeze and fluxes such as limestone and dolomite generates hazardous and greenhouse gases (GHGs). The waste gases from pellet plants include CO_2, CO, CH_4, NO_x, and SO_2. Cokemaking is also problematic from an environmental perspective as many of the hydrocarbons driven off during the coking process are additional pollutants. The coal dust generated during the transportation and charging of coal and the discharging of coke also represent significant environmental problems.

Table 3. Environmental impact of existing iron making technology

Pollutant	Cokemaking kg/met. ton iron	Sintering kg/met. ton iron	Average blast furnace kg/met. ton iron
SOx	2.02 *	1.80	17.7
NOx	0.20 *	0.67	6.81
CO	0.38 *	0.22	2.14
Particulate	3.37	0.51	5.01
VOCs	1.44 *	<0.006	0.04
CO_2	80.1	38.3	1628

Note: Calculated from data presented in several tables in a DOE Report (Energetics Inc., 2000)
* The above report states that these items do not include possibly large amounts of pollutants from various other steps for which no data are available.

All the problems associated with the iron ore agglomeration and the use of coke can be resolved only in an alternative ironmaking process that does not require them.

The intent of the Flash Ironmaking Process is to produce iron directly from fine concentrate without going through pelletization or sintering and also avoiding the need for coke.

Energy Savings

The Flash Ironmaking Process will have considerable energy benefits arising largely from the elimination of cokemaking and sintering/pelletization. Iron ore agglomeration processes such as pelletizing and sintering add considerable energy consumption to the ironmaking process. It is noted that energy consumption accompanies carbon footprint. Thus, any energy savings have corresponding reduction in carbon dioxide emissions.

Table 4. Energy balance for a commercial-scale Flash Ironmaking Process using natural gas compared with energy balance of the average blast furnace process [a]

		Flash process using natural gas	BF process
INPUT (GJ/ton of iron)	Fuel combustion [b]	13.45	8.33
	Heat recovery [c]	-4.77	-1.32
	Waste heat boiler Steam not used [d]	-3.39 -1.38	
	Sub-total	8.68	7.01
	Energy input for ore/coke preparation		5.68
	Total [h]	8.68	12.7
	Energy input for $CaCO_3$ and $MgCO_3$ calcination	0.26 [e]	
OUTPUT (GJ/ton of iron)	Reduction	0.91	2.10
	Sensible heat of iron	1.27 (1500°C)[f]	1.35 (1600°C)
	Sensible heat of slag	0.24 (1500°C)[f]	0.47 (1600°C)
	Slurry ($H_2O(l)$)	2.25 (50°C)	
	Hot water not used	1.57 (220°C)	
	Flue gas	0.79 (300°C)	0.26 (90°C)
	$CaCO_3$ decomposition		0.33
	Slag making		-0.17
	Heat loss in the reactor(s)	0.78	2.60
	Heat loss in the heat exchangers	0.73 [g]	0.07[c]
	Reactor feed gas heater Natural gas heater WGS reactor feed gas heater	0.40 0.21 0.12	(heat loss in heat recovery)
	Steam not used (90°C) [d]	0.14	
	Sub-total	8.68	7.01
	Pelletizing		3.01 [i]
	Sintering		0.65 [i]
	Cokemaking		2.02 [i]
	Total [h]	8.68	12.7
	$CaCO_3$ and $MgCO_3$ decomposition at 1000°C	0.26 [e]	

[a] Energy balance calculated by METSIM, based on the result of material balance (Table 1).

[b] Fuel combustion energy input was calculated by subtracting the energy used for iron ore reduction from the difference of heats of formation of all output components and input components. Higher heating value of natural gas was used for this calculation. If lower heating value is used, the total energy requirement decreases to 4.96 GJ of iron.

[c] Heat loss in the heat recovery step was estimated to be 5% of the energy difference between the hot offgas (calculated to be 426°C) and cooled reactor offgas (90°C).

[d] Energy recovered by steam not used in the process includes the sensible heat of steam from 300 to 90°C and the heat of condensation.

[e] The energy requirement for $CaCO_3$ and $MgCO_3$ calcination to generate CaO and MgO fluxes was separately calculated. Calcination was assumed to be done at 1000°C separately from the ironmaking process.

[f] If the temperature of iron and slag is replaced by 1600°C, the total energy requirements increase to 9.48 GJ/ton of iron.

[g] Heat losses in a heat exchanger were assumed to be 5% of the total sensible heat of input streams.

[h] Natural gas and coking coal are considered as the starting raw materials.

[i] Data obtained from (Stubbles, 2000).

An important factor considered in the new technology is the large quantities of fine iron oxide concentrates already produced in the U.S. (which do not require separate or additional comminution) that are well suited for flash reduction.

Based on detailed process simulations, a summarized comparison of energy requirements for the Flash Ironmaking Process and the average blast furnace process is presented in Table 4 (Pinegar et al., 2012).

The Flash Ironmaking Process using natural gas requires 8.7 GJ/metric ton of iron, which is 32% less than energy required by the average blast furnace process.

The higher heating value (HHV) of natural gas was used to be conservative in the computation of the energy requirement for the Flash Ironmaking Process.

The expected energy savings by the Flash Ironmaking Process using natural gas as reducing agent/fuel compared with the average blast furnace operation for annual production of 20 million tons of iron, about 40% of annual U.S. production as an example, are presented in Table 5.

For the case of operating the Flash Ironmaking Process with hydrogen, instead of natural gas, Table 6 shows a comparison of the energy requirements with the average blast furnace process.

Table 5. Expected energy savings by the application of the Flash Ironmaking Process

Flash Process (Type of Reductant /Fuel)	Flash Process (GJ/ton iron)	Blast furnace (GJ/ton iron)	Energy savings rate (GJ/ton iron)	Iron production (tons iron /year)	Annual energy savings	
					(GJ/yr)	(TBtu/yr)
	a *	B *	c = b – a	d	e = c x d	
Natural gas	8.7	12.7	4.0	$2x10^7$	80×10^6	76

* From Table 4.

Table 6. A comparison of energy requirements between the H_2-based Flash Ironmaking Process and the average blast furnace process

	H_2-based process	BF process
Total energy consumption (GJ/ton of iron)	5.48	12.7

It is seen that the Flash Ironmaking Process using hydrogen would require 5.5 GJ/ton of iron (not including energy needed for the production of hydrogen, which would vary greatly depending on production method), which is equivalent to 57% less energy consumption compared with the average blast furnace process (not including energy needed for mining, cleaning and transporting coking coal).

The results of this work indicate that the Flash Ironmaking Process would have the following significant advantages over the average current blast furnace technology:

- When natural gas is used, it will consume 32% less energy.
- When hydrogen is used, it will consume up to 57% less energy.

It is noted that a comprehensive comparison of energy requirements of the Flash Ironmaking Process and the blast furnace process is complicated and depends on the choice of starting raw materials, i.e. whether or not energy required to produce coking coal, natural gas or hydrogen is considered.

DEVELOPMENT OF FLAH IRONMAKING TECHNOLOGY

Kinetics Feasibility Tests

For accurate determination of the rate of individual iron oxide concentrate particles, a high temperature drop-tube reactor system was fabricated (Choi and Sohn, 2010). The furnace system was made up of a vertical split tube furnace with a maximum working temperature of 1540°C and a cylindrical alumina tube (5.6 cm ID, 193 cm long). The concentrate particles were injected through a tube of 0.16 cm ID carried by a hydrogen flow of 200 NmL/min. The duration of hydrogen reduction of fine iron oxide concentrate particles was determined by the residence time of particles in the isothermal zone.

The iron oxide concentrate was provided by Ternium (Monterrey, Mexico) and by ArcelorMittal. The concentrate particles are irregularly shaped. In both cases, most of the iron oxide was magnetite and the total iron content ranged from 68 to 71%. The median and mean sizes were 23.4 and 34.4 μm, respectively, and 94.4% of the total particles were less than 100 μm. The reacted powder was collected in a powder collector at the bottom of the reactor. The total iron content in the particles after reduction was determined by titration methods and the fractional reduction was obtained by tracing the difference of the mass of oxygen combined with iron in the concentrate particles before and after reduction.

Experimental data have conclusively proven that concentrate particles are sufficiently small to permit the use of a flash reduction process with its short residence time. In a drop-tube reactor, over 90% reduction was attained in 1.6 seconds and the particles were almost completely reduced in 2.5 seconds at 1200°C in large excess hydrogen, as shown in Figure 4-(a). Above 1300°C complete reduction was accomplished in less than 1.1 seconds, as shown in Figure 4-(b), and even when excess hydrogen was lowered to 40%, complete reduction was reached in less than 6.5 seconds. These results confirm that the rate is sufficiently fast for the reduction of currently available concentrate to be carried out in the flash reduction process above 1200°C.

Bench-Scale Testing

Encouraged by the promising results from the kinetic feasibility, scale-up tests have been conducted using an existing laboratory flash reactor. The flash reactor system consists of a powder feeding unit, a process gas preheater, a non-premixed burner, a reaction shaft, a powder collecting system, an off-gas scrubbing system, and various instrumentation devices. Figure 5 shows the apparatus consisting of two electrically heated parts; the upper part (0.064 m ID, 1 m long) is a hydrogen preheater and the lower part is the reactor (0.194 m ID, 1.93 m long) in which iron ore concentrate particles are reduced. They are connected by a nozzle through which hydrogen and oxygen are injected to form a H_2/O_2 flame. Hydrogen fed to the reactor is preheated to 800°C. Inside the top part of the reactor, part of the preheated hydrogen is combusted with oxygen that is supplied at room temperature. The energy released from the H_2/O_2 diffusion flame heats the hydrogen-rich reducing gas stream as well as the concentrate particles injected downward through the preheater. This hot reducing gas stream reduces the iron oxide fed at a rate of 0.1 to 1 kg/hr.

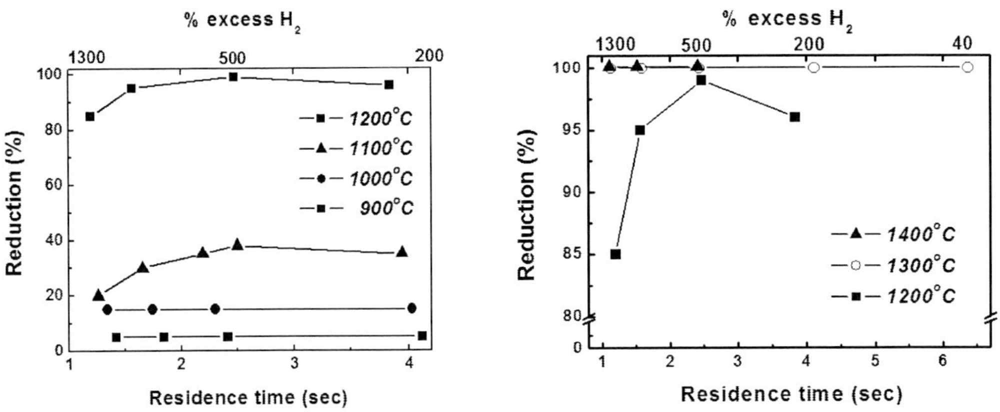

(a) 900 ~ 1200°C (b) 1200 ~ 1400 °C

Figure 4. Hydrogen reduction rate of iron oxide concentrate vs. residence time and % excess H_2 at 900 - 1400°C. (particle size: 25 - 32 μm). [Adapted from Choi and Sohn (2010)]

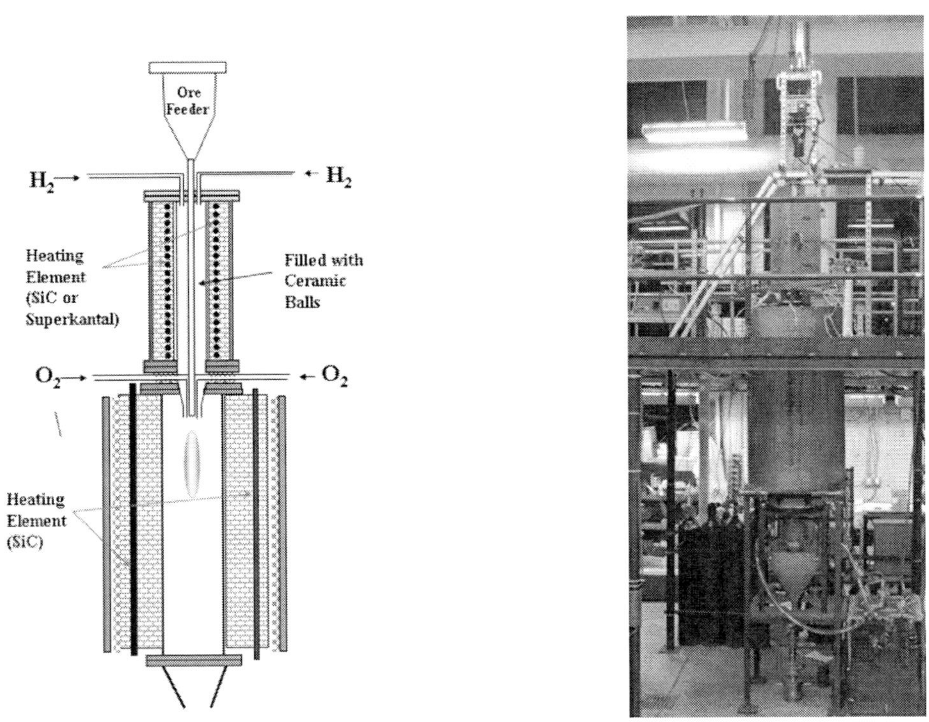

(a) (b)

Figure 5. Laboratory flash reactor: (a) schematic diagram; (b) photograph.

Heating elements around the main reactor supplies additional energy provides partial compensation for heat loss, and also keeps the H2/O2 diffusion flame more stable by maintaining the reactor above the auto-ignition temperature (about 570oC for a stoichiometric mixture at normal atmospheric pressure). The off-gas stream consisting of particles and gas passes through a dust collecting chamber in which its velocity decreases, thus allowing the particles to be separated from the main gaseous stream. The reduced particles fall into the receptacle.

Table 7. Chemical composition of the iron ore concentrates from ArcelorMittal

Component	Total Iron	FeO	SiO_2	CaO	MgO	Al_2O_3	TiO_2	MnO	Na_2O	K_2O	P	S
wt%	68.22	28.24	4.30	0.19	0.29	0.061	0.021	0.021	0.011	0.044	0.014	0.0005

Table 8. Three Representative Cases

Sample	RD, %	Powder Feeding Method	Sampling Location
1	100	Pneumatic [1]	Off-gas passage
2	72	Pneumatic [1]	Bottom reactor flange
3	35	Gravity[2]	Collecting chamber

[1] H_2 used as a carrier gas.

[2] Disc-type micro-feeder. Powder fed through the burner.

The concentrate used for the experiments was provided by ArcelorMittal (East Chicago, USA) and had the composition presented in Table 7. The reduced concentrate was sampled for chemical analysis at three different locations: the powder collecting chamber, the bottom flange of the reactor, and the elbow connections. The samples were then characterized by various analytical techniques and the degree of reduction is determined.

Table 8 presents the most representative product samples collected at different parts of the reactor from different experimental runs. Also, the powder feeding method used in the corresponding experiment is indicated. In case of the gravity feeding, the powder was supplied using a disc-type micro-feeder. The feeding port was located at the center of the burner. As a result, the concentrate powder went through the flame. On the other hand, the pneumatic feeding was carried out through an auxiliary feeding port located between the hydrogen flame and the inner wall of the reactor.

Sample 1 represents the fine particles that are entrained in the flow of the process gas and exit the system as dust. They were collected from the elbow connections used in the off-gas system. SEM micrographs, Figure 6-(a), indicated that the largest particles have sizes smaller than 20 μm. XRD results, Figure 7-(a), indicated that the predominant phase is metallic iron.

Sample 2 was collected from the reactor bottom flange. SEM micrographs, Figure 6-(b), show that the particles almost kept the original shape of the concentrate but they lost the sharp edges and developed porosity. XRD results, Figure 7-(b), suggest that the metallic iron is the most abundant phase. The chemical analysis by titration indicated that the reduction degree of this sample was 72%.

In the last case (Sample 3), the iron concentrate was fed using a disc-type micro-feeder. SEM micrographs, Figure 6-(c), showed spherical porous particles with sizes larger than the average size of the concentrate powder. This is explained by the fact the particles melted upon contacting the flame before undergoing reduction. XRD results, Figure 7-(c), showed that metallic iron, wustite and magnetite phases are present in the collected powder. This result is consistent with the chemical analysis that indicates a reduction degree of 35%. EDS results indicated that the particle surface is mostly composed of metallic iron.

Due to the limitations inherent in its design and reactor body material, testing in the laboratory flash reactor is not able to provide all the necessary information to assess the Flash Ironmaking process as an industrially viable technology: the major limitation in the current laboratory flash reactor is to provide a sufficient amount of heating. Thus, a large-scale bench reactor test program is in progress.

(a) (b) (c)

Figure 6. SEM micrographs of Samples 1~3 with reduction degrees of (a) 100%; (b) 72%; (c) 35%.

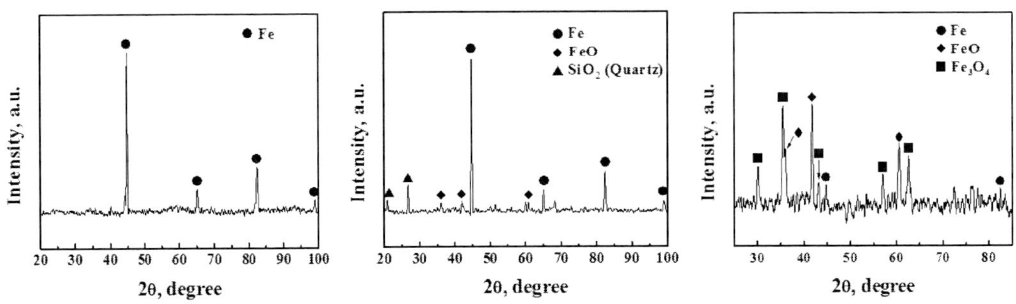

(a) (b) (c)

Figure 7. XRD patterns of Sample 1~3 with reduction degrees of (a) 100%; (b) 72%; (c) 35%.

In identifying the specific design and functional features necessary for the bench facility, the method of supplying heat has been the key issue. This bench reactor will produce iron using hydrogen or natural gas as the fuel as well as the reductant in a cylindrical reactor. The energy for the sensible heat of solid and gas feed materials and for the reduction reaction of iron oxide will be supplied by a single burner installed on top of the reduction chamber in combination with a plasma torch as a supplementary heat source.

Figure 8 shows the conceptual diagram and preliminary facility layout of the bench-scale test facility comprising several subsystems such as a hydrogen burner, a pneumatic powder feeder, gas delivery lines, safety measures and interlocking system, a quenching chamber, a powder collection bin, and a flare stack to burn the hydrogen remaining in the off-gas. (In an industrial plant, the off-gas will be recycled after removing water vapor contained in it.) A gas burner will generate a flame inside the top part of the reaction chamber (0.81 m ID, 1.52 m high). The design feed rate of iron ore concentrate in this scale is 5 kg/h. Concentrate particles will be supplied through the burner or in the vicinity of the hydrogen/oxygen flame. The flame will also be the main energy source for the sensible heat for the process. The average temperature of the reaction shaft will be 1400~1600°C. The entire system will be monitored by several safety measures and controlled by proper emergency shut-off procedures.

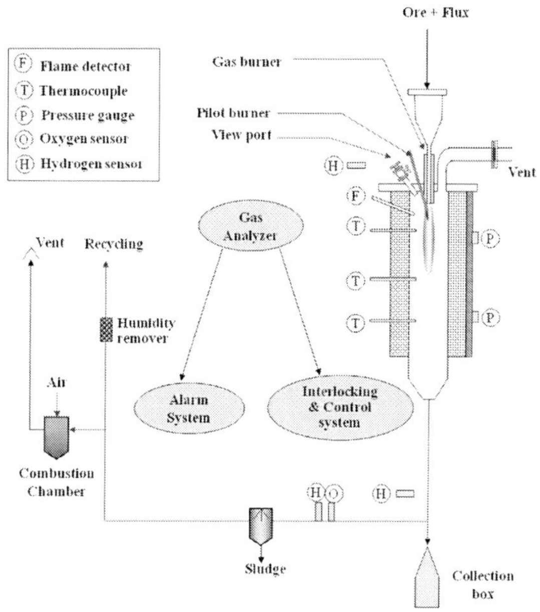

(a)

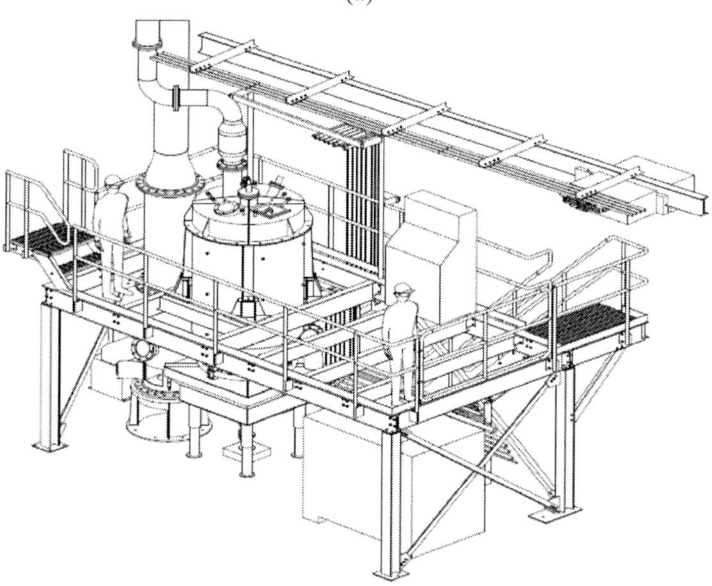

(b)

Figure 8. Large-scale bench reactor: (a) conceptual diagram; (b) preliminary facility layout.

It is noted that the bench reactor has many features of an eventual industrial reactor. They include a burner as the main heat source and a refractory-lined vessel. Upon completing this project phase, engineering data necessary for industrial pilot-plant operations will have been generated, such as heat loss, product collection, raw materials feeding, process control strategy, burner design, and off-gas treatment.

CONCLUDING REMARKS

An entirely new transformational process for alternate ironmaking is under development based on the direct gaseous reduction of iron oxide concentrates in a flash reduction process. The novel Flash Ironmaking Technology will significantly lower carbon dioxide emissions by 61-96% and reduce energy consumption by 32-57%, depending on the type of the fuel/reductant used, compared with the average blast furnace ironmaking route. The process uses gaseous reducing agents, such as natural gas, hydrogen, other syngas, or a combination thereof. It is to be applied to the production of iron as a feed to the steelmaking process, eventually replacing the blast furnace and other alternate ironmaking processes. The process can also be a part of a continuous direct steelmaking process.

Before an industrially viable technology based on the new concept can be developed, bench- and pilot-scale tests must be carried out to identify technical hurdles and develop methods to overcome them until a technically and energy-wise superior industrial process can be achieved. Justified by experimental data obtained during the kinetics feasibility tests, a scale-up development work is currently on-going at the University of Utah. Testing in a laboratory flash furnace has resulted in the establishment of a kinetics database over wide ranges of operating conditions and a complete design of a more advanced bench reactor. With an objective to develop an industrially viable Flash Ironmaking Technology, a comprehensive bench-scale testing campaign is planned.

The deliverables are expected to be the determination of the scalability of the process, substantive process simulation results and fundamental engineering data leading to the design and construction of an industrial pilot plant.

REFERENCES

Camlic, R. L, Goodman, N. J., and Kvaerner, D., 1997, Upgrade of process control system at U.S. Steel Fairfield No.8 blast furnace, *Ironmaking Conference Proceedings*, 56, pp. 145-154.

Chang-qing, H. U. et al., 2006, Emission Mitigation of CO2 in Steel Industry - Current Status and Future Scenarios, *Journal of Iron and Steel Research International*, 13 (6), pp. 38-42.

Choi, M. E. and Sohn, H. Y., 2010, Development of a Green Suspension Ironmaking Technology Based on Hydrogen Reduction of Iron Oxide Concentrate: Rate Measurements, *Ironmaking Steelmaking*, 37, pp. 81-88.

Duarte, P. E. et al., 2010, Energiron Direct Reduction Ironmaking – Economical, Flexible, Environmentally Friendly. *Steel Times International*, p. 25.

Emi, T. and Seetharaman, S., 2000, Future Steelmaking Plant with Minimized Energy Consumption and Waste Evolution, *Scandinavian Journal of Metallurgy*, 29, pp. 185-193.

Energetics Inc., 2000, *Energy and Environmental Profile of the U.S. Iron and Steel Industry,* U.S.Department of Energy, (online) Available at: <http://www1.eere.energy.gov/industry/steel/pdfs/steel_profile.pdf>(Accessed 23 Jan 2012).

Gielen, D., 2003, CO_2 Removal in the Iron and Steel Industry, *Energy Conversion and Management*, 44, pp. 1027-1037.

Karl, T. R., Melillo, J. M. and Peterson T. C., 2009, *Global Climate Change Impacts in the Unites States*, Cambridge, NY: Cambridge University Press, 2009.

Kim, Y. and Worrell, E., 2002, International Comparison of CO_2 Emission Trends in the Iron and Steel Industry, *Energy Policy*, 30, pp. 827-838.

Lu, W. –K., 1997, The Impact of the Environmental Concerns on the Development of Ironmaking Technology, *Acta Metallurgica Sinica*, 33 (2), pp. 207-212.

Pinegar, H. K., Moats, M. S., Sohn, H. Y., 2011, Process Simulation and Economic Feasibility Analysis for a Hydrogen-Based Novel Suspension Ironmaking Technology, *Steel Res. Int'l.*, 82, (8), pp. 951–963.

Pinegar, H. K., Moats, M. S., and Sohn, H. Y., 2012, Flow Sheet Development, Process Simulation and Economic Feasibility Analysis for a Novel Suspension Ironmaking Technology Based on Natural Gas: Part I. Flow Sheet and Simulation for Ironmaking with Reformerless Natural Gas; Part II. Flow Sheets and Simulation for Ironmaking Combined with Steam-Methane Reforming; Part III: Economic Feasibility Analysis, *Ironmaking Steelmaking*, in press.

Schwartzel, E., 2011, Penn State forecasts boom for Marcellus Shale, Pittsburgh Post-Gazette, (online) Available at: <http://www.post-gazette.com/pg/11202/1161933-503.stm> (Accessed 23 Jan 2012).

Sohn et al., 2009, Suspension Reduction Technology for Ironmaking with Low CO_2 Emission and Energy Requirement, *Iron and Steel Technology (AIST Trans.)*, 6 (6), pp. 158-165.

Stubbles, J., 2000, Energy Use in the U.S. Steel Industry: An Historical Perspective and Future Opportunities, U.S. Department of Energy, (online) Available at <www1.eere.energy.gov/industry/steel/pdfs/steel_ energy_use.pdf> (Accessed 23 Jan 2012).

Tacke, K. –H. and Steffen, R., 2004, Hydrogen for the Reduction of Iron Ores - State of the Art and Future Aspects, *Stahl und Eisen*, 124 (4), pp. 45-52.

USGS (U.S. Geological Survey), 2005, Minerals Yearbook - Iron Ore, (online) Available at:<http://minerals.er.usgs.gov/minerals/pubs/commodity/iron_ ore/> (Accessed 23 Jan 2012).

USEIA (U.S. Energy Information Administration), 2011a, Annual Energy Outlook, (online) Available at:< http://www.eia.gov/forecasts/aeo/index. cfm> (Accessed 23 Jan 2012).

USEIA (U.S. Energy Information Administration), 2011b, Review of Emerging Resources: U.S. Shale Gas and Shale Oil Plays, (online) Available at:<http://www.eia.gov/analysis/studies/usshalegas/> (Accessed 23 Jan 2012).

U.S. Steel Corp., 1985, *The making, shaping and treating of steel* 10th edition, Association of Iron and Steel Engineers.

Wang, H., 2011, Reduction Kinetics of Iron Ore Concentrate Particles Relevant to a Novel Green Ironmaking Process, *Ph.D. Dissertation,* University of Utah, Salt Lake City, Utah.

In: Carbon Dioxide Emissions: New Research
Editors: M. Carpenter and E. J. Shelton

ISBN: 978-1-62257-436-0

Chapter 4

CO_2–E EMISSIONS FROM BEST PRACTICE LANDFILL VERSUS COMPOST

Barbara Hutton[1], Edmund Horan[1] and Mark Norrish[2]
[1]Sustainable Practice Program, RMIT University, Melbourne, Australia
[2]Computational Mathematics ANU, Canberra, Australia

ABSTRACT

Carbon dioxide (CO_2), methane (CH_4) and nitrous oxide (N_2O) are together major greenhouse gas emissions from waste management practices. Methane and nitrous oxide are predicted to cause as much global warming in terms of CO_2 equivalents (CO_2e) as carbon dioxide over the next 20 years, according to the Intergovernmental Panel on Climate Change. Methane's global warming impact (GWI) over 100 years is 25 times that of CO_2. This impact is higher in the first 20 years. N_2O is even more potent, with 298 times the GWI of carbon dioxide, based findings by the Intergovernmental Panel on Climate Change (IPCC 2007).

It is well-known that methane is produced in anaerobic digestion, as in landfill. It is less well-known that methane and N_2O are both emitted in small but significant quantities when organic material is composted (IPCC 2006).

The City of Melbourne, Australia has strict Environment Protection Agency regulations controlling landfills. Best practice landfills have measures such as vacuum gas extraction which can capture greater than 75% of the methane produced (NGERS Technical Guidelines 2010), and this methane is used in the generation of electricity, thus reducing CO_2 emissions from coal-fired energy generation. Yet Melbourne also has a policy of diversion of municipal waste to compost, which reduces this yield of landfill gas.

The State of Victoria set a target for diverting wastes from landfill by 65%, between 2005 and 2014. Victoria's "Towards Zero Waste Strategy Report 2010" states that this policy "reduces energy use and carbon emissions and helps mitigate climate change". Is this true? This study compares expected CO_2-e emissions from landfill (with methane capture) versus emissions from municipal composting, using local data on Melbourne's waste composition. Calculations were based on Australian guidelines for estimating CO_2-e emissions (DCCEEE 2011) and IPCC (2006).

Findings: Cumulative CO_2-e emissions in Victoria are in fact significantly lower when organic material is sent to best practice landfill sites, rather than to well-turned municipal compost sites.

Discussion

With current technology approximately 78% of methane produced in landfill can be captured (DCCEEE Technical guidelines 2010, GHD 2010). By comparison, compost emits small amounts of methane as well as N_2O, a very potent greenhouse gas. N_2O is not formed in landfill due to lack of oxygen: anaerobic digestion produces N_2, which is not a greenhouse gas (Otte et al 1996).

Landfill emits a lower volume of CO_2 than aerobic composting. This is because approximately half of the organic carbon in anaerobic landfill is sequestered: it does not decompose due to lack of oxygen (IPCC 2006)). A further half of the carbon that does decompose is taken up as methane.

The additional CO_2 produced by composting is one significant factor in explaining why CO_2-e emissions from municipal compost are higher than those from landfill – but even if CO_2 is excluded, municipal compost still has higher total CO_2-e emissions, viewed over 40 years.

This study measured only direct emissions from best practice landfill, compared with those from frequently turned municipal compost. It did not factor in the further CO_2-e savings made when biogas from landfill is used as a fuel to substitute for brown coal in energy generation.

Home composting is likely to produce more methane and N_2O per tonne of waste treated than municipal composting, because of the use of plastic compost bins. These are more difficult to aerate than municipal windrow compost, which is mechanically turned on a regular basis. Diversion of organics to home composting could be particularly counterproductive from a CO_2-e perspective.

1. Background

The Intergovernmental Panel on Climate Change (IPCC) is an international panel of experts established in 1988 by the United Nations Environment Programme (UNEP) and the World Meteorological Organization (WMO) to provide the world with a clear scientific view on the current state of knowledge about climate. Thousands of leading scientists worldwide participate, and their work is subject to continuous review. It provides 194 IPCC member nations with guidelines for assessing their greenhouse emissions. These guidelines are regularly updated as more scientific information becomes available.

According to the IPCC, the three main greenhouse gases are carbon dioxide, methane and N_2O (commonly called nitrous oxide). Methane (CH_4) and N_2O have a higher global warming potential than CO_2. Global warming potential is often expressed in terms of carbon dioxide equivalents or CO_2-e.

Methane has 25 times the global warming impact of CO_2 (measured over 100 years), according to IPCC Guidelines, 2006. Initially a figure of 21 was used but this has been updated due to new scientific information. IPCC 2007 also explains that while CO_2 survives for about 100 years in the atmosphere, methane breaks down relatively quickly, hence it is more meaningful to look at it over a 20 year span. Over this period methane has a carbon dioxide equivalence of 72, while N_2O's is 298 times that of CO_2.

When organic material decomposes in an airless situation, as in landfill, it is described ~~an~~ as anaerobic digestion, and a mixture largely consisting of CO_2, CH_4 and water is produced.

Due to the high global warming potential of methane, anaerobic decomposition is undesirable unless methane emissions can be captured. In recent years, however, most developed countries including Australia have taken action to control greenhouse emissions from landfill. The EPA in Victoria provides a code of "Best Practice" for landfill (EPA Victoria 2010). Major Victorian solid waste disposal sites are now required to install covers and vacuum gas collection systems to capture the methane, which is used to generate renewable energy. This practice is widespread in the developed world. The proportion of methane captured is difficult to measure but it is assumed that at least 75% can be captured in "best practice" landfill systems (DCCEE Technical guidelines 2010).

The 2006 version of the IPCC Guidelines includes revised assumptions on the decomposition of organic waste in landfill, and a new section on emissions from compost. It incorporates findings that organic waste in anaerobic landfill breaks down slowly, and approximately 50% never breaks down at all; thus 50% less methane is generated than previously thought. Earlier guidelines, which assumed high initial production of landfill gas, appeared to show that much of the methane generated must have been escaping from landfill. The revised guidelines suggest this is not the case.

The guidelines also make explicit for the first time that even well-aerated municipal compost windrows, in airy outdoor conditions, tend to emit small but significant quantities of methane and N_2O. Good aeration can reduce but not prevent emissions. Nine studies on this topic are listed (IPCC 2006, Vol 5, p 4.8). One of these studies, Hellebrand (1998), detected methane in supposedly aerobic compost heaps when green matter was compacted at the bottom. N_2O was generated at low, but not anaerobic, oxygen levels.

A study by Amlinger et al (2008) measured methane levels when compost is turned. Turning releases trapped methane and heat, caused by decomposition. High temperatures tend to discourage microbial activity, so turning the compost and cooling it encourages renewed microbial activity, and soon methane levels in the compost rebound.

Table 4.1 of the IPCC (2006) guidelines shows that, per kilogram of wet waste, aerobic compost emits approximately:

- 4 grams of methane (range: 0.03 – 8 grams) and
- 0.3 grams of N_2O (range: 0.06 - 0.6 grams).

National governments including Australia use these values when estimating their greenhouse emissions from municipal composting (DCCEE Technical Guidelines 2010). In poorly aerated waste (e.g. domestic compost in a plastic bin) or in high-nitrogen wastes such as manure, N_2O emissions can be far higher than the IPCC (2006) default value, according to both Amlinger (2008) and Hellebrand (1998).

2. State and Municipal Government Policies on Landfill and Compost

Victoria has an ambitious target to divert waste from landfill. Metals and paper recycling provide significant revenue to offset the collection costs. Recycling of "comingled waste" (metals, glass, paper, plastic) saves significant amounts of resources, water, energy and greenhouse gas (Centre for Design 2001).

But in Victoria, the level of comingled recycling has in fact fallen over time, to be replaced by an emphasis on organics recycling. Organic waste diversion increased by 13% in 2009-10, to 924,000 tonnes of the 1.5 million tonnes of waste recovered from landfill, while at the same time paper and cardboard recycling fell.

The State of Victoria has set a target for diverting wastes from landfill by 65%, between 2005 and 2014. This policy does not discriminate between types of waste: organic or comingled. Victoria's "Towards Zero Waste Strategy Report 2010" states that this policy "reduces energy use and carbon emissions and helps mitigate climate change". No evidence for this claim is cited.

Municipal governments, which are responsible for municipal waste management, have policies based on the State waste diversion targets.

Most popular literature on composting assumes that aerobic compost does not produce methane. Some sources add the proviso that it should be turned regularly to aerate it, but this is rarely mentioned in municipal council websites. Most municipal councils in Melbourne (Victoria) have instituted green waste and household organic collections to divert organic waste from landfill, and promote home composting or worm farms, in the belief that this will reduce CO_2-e emissions. No references are cited, and the information is garbled in some sites.

For example, Mornington Peninsula Shire Council's website states:

"Decomposing food and garden waste in landfill can generate methane gas – a potent greenhouse gas... By composting at home, it is estimated that the average household can save over 1,600 black balloons or 800kg of greenhouse gas emissions per year."

Another Council website states: "Organic material goes to landfill where it breaks down without oxygen and produces leachate, a toxic pollutant greenhouse gas that is 21 times more potent than carbon dioxide."

This compost bin, provided by Knox council to residents, is typical of the bins offered to residents. Note the lack of air holes, and the difficulty of turning waste material inside the bin. This would encourage production of anaerobic compost, emitting N_2O and methane.

Government waste management policies may be exacerbating CO_2-e emissions rather than reducing them. Funding for these campaigns is provided by taxpayers and ratepayers, or diverted from other, perhaps more worthwhile, campaigns.

3. Aim of This Study

This study aims to compare CO_2-e emissions from 1 tonne of domestic organic waste material, deposited annually in landfill, versus emissions from 1 tonne per year treated in municipal windrow composting. The study considers only direct emissions of greenhouse gases (N_2O and methane have been modelled in detail).

It does not take into consideration CO_2 emissions from transport and treatment, which are expected to be higher for composting, nor CO_2 emissions from coal-fired electricity generation avoided by use of landfill biogas to produce power.

4. Method and Assumptions

This study uses the IPCC Guidelines for National Greenhouse Gas Inventories (2006), which have been adopted by the Australian Government (DCCEE 2011 and 2010), to compare total CO_2-e emissions from best practice landfill in Melbourne (DCCEE 2011 chapter 5), with expected emissions from municipal composting (DCCEE 2011 page 49). It assumes that the municipal compost is turned regularly in an open-air location.

Information on the breakdown of waste types in Melbourne's landfill is derived from a Victorian Department of Sustainability and Environment survey conducted in 2008, which found that household garbage then contained the following organic matter: 41% food waste, 7.2% garden waste, 12.1% total paper and 3.1% nappies. (DSE 2008). Rates of breakdown for the different types of waste are provided in (DCCEE 2011 chapter 5).

A spreadsheet was set up to compare emissions of methane, nitrous oxide and the resultant CO_2-e emissions from compost, landfill and incineration. The IPCC model allows for differences in temperature, humidity, dryness and aeration in the landfill, different types of organic waste, and variations between temperate and tropical climates. Landfill in Melbourne (Australia) was used as a case study for the spreadsheet.

Information on local rates of methane capture was provided by staff at Western and Wollert Landfills in outer Melbourne. A study of Wollert landfill (GHD 2009) found that the rate of methane capture was between 70 and 85% during active management of the site, while the Department of Climate Change (DCCEE Technical Guidelines 2010) assumes that methane capture from landfill of greater than 75% is possible. We therefore assumed 78% of landfill gas is captured.

Under IPCC 2006 Guidelines, it is assumed that 10% of escaping methane from landfill is oxidised as it passes through the soil cover. Melbourne landfills have soil cover, so we adopted this value, along with default values for emissions from landfill and composting.

The CO_2-e savings achieved by using landfill biogas to produce electricity, in lieu of brown coal, the main fuel for electricity generation in Victoria, were not taken into account in these calculations. Including reduction in emissions from brown coal would have produced a more favourable outcome for landfill in comparison to compost, which is not used to produce electricity.

Diesel use is likely to be higher for composting than landfill, because compost must be regularly turned by diesel-operated vehicles and because more transport is involved. Landfill is taken to one outer suburban location and buried; compost goes to another outer suburban composting site, then it is transported again e.g. to a gardening products supplier, before its final journey to a garden. Again this likely variation in diesel use was not factored in.

Using compost in lieu of manufactured nitrogen fertiliser could theoretically save CO_2-e emissions because man-made nitrogen fertiliser is fossil-fuel-intensive to manufacture. A preliminary analysis suggested that the savings are probably small or negative, as manufactured fertiliser it is compact and easy to apply, whereas compost requires more energy to transport and spread. In the Australian context, with large distances between urban centres and farms, this would outweigh the energy savings in manufacture. Furthermore, Melbourne's household organics are not in fact used as agricultural compost (at time of writing and researching this report). This is largely for economic reasons such as transport costs. Green waste is used only in potting mix. In summary these calculations simply examine gross greenhouse emissions from each decomposition process.

5. Findings

Over 40 years, CO_2-e emissions from organic waste treated at best practice landfill sites were in fact lower than those from well-turned aerobic compost.

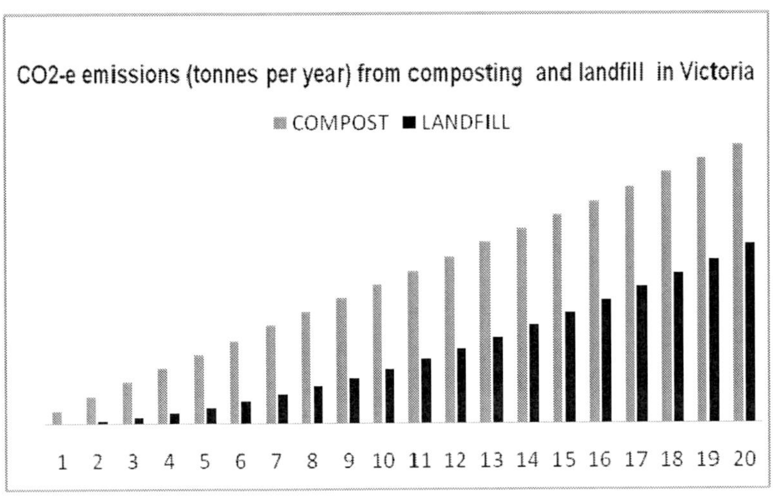

Figure 1. CO_2-e emissions from compost versus landfill in Victoria, over 20 years. The gases considered here are methane and N_2O, converted to CO_2-equivalent.

After 15 years, emissions from the waste treated in landfill were just over half those from municipal composting and they were still lower after 40 years.

6. Discussion

(i) Compost

It is not well-known that aerobic composting emits methane. Even if compost is turned regularly, methane forms inside moisture-coated particles of material. Compost also emits small but significant amounts of N_2O: due to its high GWP this gas is responsible for almost half the anthropogenic CO_2.e emissions from compost under optimal conditions. In poorly aerated waste, e.g domestic compost in plastic bins, manure or sewage sludge, N_2O emissions can be far higher (Hellebrand 1998, Amlinger 2008).

The following table (from Amlinger, 2008) shows the proportion of methane in emissions from backyard composting, and from three types of windrow compost (windrows are mechanically-aerated open piles – this is how municipal compost is traditionally treated).

Biowaste including kitchen waste produced a higher proportion of methane than green waste, with backyard composting producing the highest methane count of all types. Dr Florian Amlinger, an exponent of composting, made the backyard compost in a manner that would allow maximum aeration, by alternating food scraps with layers of woodchips in a wooden box with slats. Backyard compost made of solid food and green-waste in a plastic bin is likely to have still higher methane emissions.

Table 1. (condensed from Amlinger 2008): Proportion of methane in emissions from three types of windrow compost, and from backyard composting

Windrow composting turned 1 to 2 times weekly	Grams CH_4/kg CO_2
Biowaste (including kitchen waste)	5.83 - 18.77
Green waste	1.13 - 8.58
Sewage sludge	4.84 - 10.08
Backyard composting	15.59 – 27.94

The above table shows that the backyard compost emits 3 – 12 times as much methane as the municipal green waste.

(ii) Landfill

Due to lack of oxygen, N_2O is not formed in anaerobic landfill: nitrogen is emitted as N_2, which is not a greenhouse gas (Otte et al 1996).

The IPCC does not require member states to include CO_2 emissions from waste in their national greenhouse reporting as they are not considered to be anthropogenic, and they are not factored into the results shown in Figure 1. In a study comparing total CO_2-e emissions from two waste treatments, CO_2 itself cannot be completely ignored, however.

Landfill emits only about one quarter as much CO_2 as composting, per tonne of waste material. This is because approximately half of the organic carbon in anaerobic landfill is sequestered: it does not decompose due to lack of oxygen (IPCC 2006)). A further half of the carbon that does decompose is taken up as methane. CO_2 has a low GWP of 1, compared with 25 for methane and 298 for N_2O, however it is heavier than methane. When estimating greenhouse gas emissions, the amount of gas by weight is multiplied by the GWP.

For example, carbon has a relative atomic mass of approximately 12, hydrogen 1 and oxygen 16.

CO_2 thus has an atomic mass of 44 compared to 16 for CH_4 (methane). CO_2 is emitted in large quantities during aerobic composting.

The additional CO_2 produced by composting is one significant factor in explaining why long-term CO_2-e emissions from municipal compost are higher than those from landfill – but even if CO_2 is excluded, municipal compost still has higher total CO_2-e emissions, when viewed over 20 years, as shown in Figure 1, or 40 years (not shown).

This study measured only direct benefits from reducing CH_4 and N_2O emissions from composting. If energy-from-waste is considered, results for incineration and landfill gas capture are even better.

Victoria has an ambitious target to divert waste from landfill but does not distinguish between diversion of wastes that would minimise CO_2-e emissions and those that do not. Organic waste diversion increased by 13% in 2009-10, to 924,000 tonnes of the 1.5 million tonnes of waste recovered from landfill, while at the same time paper and cardboard recycling fell. Metals and paper recycling provide significant revenue to offset collection costs, and, with glass, save significant amounts of resources, energy and greenhouse gas (Centre for Design 2001).

This is not the case with diversion of organic material. Composting emits more CO_2-e, and also brings the emissions forward. The CO_2-e is produced within a few weeks in composting, when in landfill it takes 20 years or more. Thus with composting climate change may be accelerated.

Emissions from landfill may continue at a declining rate for years after the closure of landfill, but in future, methane capture technology is likely to be improved. Old landfill also may be aerated by mining of landfilled metals; methane may be oxidised in cover material or offset by growth of plants on former landfill sites. Diverting organic waste to compost now, without capturing emissions from the compost, may be counter-productive.

CONCLUSION

Resources devoted to diverting organic waste from landfill could be better spent on ensuring all landfill does have best practice methane capture; addressing emissions from manures, sewage sludge or increasing recycling of packaging waste (Grant et al 2001).

A study by Pipatti and Savolainen (1996) and the Hyder Study of waste options for Melbourne (2007). both found that incineration was in fact the best option for reducing greenhouse gases from a CO_2-e perspective. This area requires further research.

The above two studies found that anaerobic digestion of wastes for methane would work well if 100% of organic waste could be separated, but this is unlikely to happen in reality. Pipatti and Savolainen found that the best option for CO_2-e reduction, other than perhaps incineration, was landfill treatment of organic waste, with CH_4 captured and used as fuel. Our study supports this finding.

Melbourne has adequate landfill space, and more is created by quarrying activities. The huge volume of poor compost produced if all household waste is composted may lead to a collapse in the market for compost. Well managed landfill with gas capture can reduce methane levels and delay emissions for decades. Spark ignition motors are currently used to convert methane to electricity, but fuel cells, cogeneration of energy and heat, or cleaning of methane and direct use as fuel are all possible.

APPENDIX 1. FACTORS USED IN CALCULATION OF CO_2-E EMISSIONS FROM LANDFILL VERSUS MUNICIPAL COMPOSTING

Calculations of emissions from Landfill are based on formulas in DCCEE 2011 National Greenhouse and Energy Reporting (Measurement) Technical Guidelines, Chapter 5 – *Waste: Solid waste disposal on land 5 part 5.2*

Calculations of emissions from compost are based on DCCEE (2011): Department of Climate Change and Energy Efficiency *National Greenhouse Accounts Factors July 2011*, p. 49, "Biological Treatment of Wastes" and the slightly more detailed *2006 IPCC Guidelines for National Greenhouse Gas Inventories, Vol 5, Chs 4, table 4.1.*

Landfill:	
Variables related to waste	kg
Quantity of organic waste treated:	1.00
Fraction of degradable organic carbon (wet weight)	0.20
Fraction of DOC that degrades in anaerobic site	0.50
MCF methane correction factor	1.00
Therefore the mass of decomposable org. carbon that degrades:	0.10
Variables related to time and decay rate	
k_j = decay rate for waste type j	0.052
e $-k_j$ rate of accumulation of waste	0.948
Conversion to methane	
Ratio CH_4:C = 16:12	1.333
Fraction of CH_4 in gas	0.5
Variables related to SWDS	
(1-r) R = recovered. Assume 0.78	0.22
(1-OX) OX = amount oxidised in cover, assume	0.10
	0.90
The above 4 factors multiplied together	0.1320
Compost:	
Methane emissions:	
Mass of organic waste treated, kg	1.00
EF emission factor for composting	0.004 kg/kg waste
GWP=21	0.084
Nitrous oxide emissions	
EF emission factor for composting	0.0003 kg/kg waste
GWP=298	0.089
Total CO2-e from methane and N_20	0.173 Kg CO_2-e

LANDFILL: CO2-E EMISSIONS OVER 20 YEARS

	A	B	C	D	E	F
Year	Accumulated Decomposable waste	Decomposed waste	Methane emitted yr y	Cumulative methane	CO_2-e from compost	CO_2-e from landfill
1	0.100	0.005	0.00	0.00	0.17	0.014
2	0.195	0.010	0.00	0.00	0.35	0.042
3	0.285	0.015	0.00	0.00	0.52	0.084
4	0.370	0.019	0.00	0.01	0.69	0.137
5	0.451	0.023	0.00	0.01	0.87	0.202
6	0.528	0.027	0.00	0.01	1.04	0.278
7	0.600	0.031	0.00	0.02	1.21	0.364
8	0.669	0.035	0.00	0.02	1.39	0.461
9	0.735	0.038	0.01	0.03	1.56	0.567
10	0.797	0.041	0.01	0.03	1.73	0.682
11	0.856	0.045	0.01	0.04	1.91	0.805
12	0.912	0.047	0.01	0.04	2.08	0.936
13	0.965	0.050	0.01	0.05	2.25	1.075
14	1.015	0.053	0.01	0.06	2.43	1.222
15	1.062	0.055	0.01	0.07	2.60	1.375
16	1.108	0.058	0.01	0.07	2.77	1.534
17	1.150	0.060	0.01	0.08	2.95	1.700
18	1.191	0.062	0.01	0.09	3.12	1.872
19	1.230	0.064	0.01	0.10	3.29	2.049
20	1.266	0.066	0.01	0.11	3.47	2.231

Column A: Cumulative organic carbon in landfill at end of Year y: (decomposable waste for year y + 0.948 x cumulative total); Column B: Decomposable waste x decay rate, 0.52 p.a., Column C: Methane actually emitted to the atmosphere in year y.; Column D Cumulative emissions from landfill; Column E: Cumulative emissions from municipal compost.

REFERENCES

Amlinger, F, Peyr, S and Cuhls, C (2008) "Greenhouse gas emissions from composting and mechanical biological treatment" *Waste Management Research* 26: 47.

DCCEE, Department of Climate Change and Energy Efficiency (2011) *National Greenhouse Accounts Factors July 2011*, p. 49, "Biological Treatment of Wastes". http://www.climatechange.gov.au/~/media/ publications/greenhouse-acctg/national-greenhouse-accounts-factors-july-2011.pdf.

DCCEE, Department of Climate Change and Energy Efficiency (2010) *National Greenhouse and Energy Reporting (Measurement) Technical Guidelines June 2010*, Chapter 5 – Part 5.2 "http://www. climatechange.gov.au/~/media/publications/greenhouse-acctg/national-greenhouse-accounts-factors-july-2011.pdf.

Department of Sustainability and Environment (Vic), 2008: "Kerbside Garbage Composition: Recent Findings". (Internet search: Kerbside_Garbage_Composition_Recent_Findings21).

EPA Victoria 2010: *Best Practice Environmental Management: Siting, Design, Operation and Rehabilitation of Landfills* Publication 788.1 September 2010.

GHD October 2010 "Report for Wollert Landfill: Flux Testing and Emissions Estimates", Hanson Landfill Services, Melbourne.

Grant, Tim, Karli James, Sven Lundie and Kees Sonneveld, 2001.

"Stage 2 Report for Life Cycle Assessment Paper and Packaging Waste Management in Victoria" For EcoRecycle Victoria, by CRC for Waste Management and Pollution Control; Centre for Design, RMIT; Centre for Waste and Water Technology, UNSW; Centre for Packaging, Transportation and Storage at VU; Food and Packaging CRC.

Hellebrand, H.J. 1998. "Emission of nitrous oxide and other trace gases during composting of grass and green waste", *Journal of Agricultural Engineering Research.* 69: 365-375.

IPCC (2006a) Guidelines for National Greenhouse Gas Inventories, Waste Vol 5, Ch 3, p 3.14. http://www.ipcc-nggip.iges.or.jp/public/2006gl/pdf/ 5_Volume5/V5_3_Ch3_SWDS.pdf.

IPCC (2007) Changes in Atmospheric Constituents and in Radiative Forcing, p 212. http://www.ipcc.ch/pdf/assessment-report/ar4/wg1/ar4-wg1-chapter2.pdf.

Otte, S., Grobben, N., Robertson, L., Mike S. M., Jetten, M. and. Gijs Kuenen, J, 1996: "Nitrous Oxide Production by Alcaligenes faecalis under Transient and Dynamic Aerobic and Anaerobic Conditions", *Applied and Environmental Microbiology*, July 1996, p. 2421–2426 Vol. 62, No. 7.

Municipal council websites: Mornington: www. mornpen.vic.gov.au/ Page/page.asp?Page_Id=1279andh=0. Frankston: www.frankston.vic. gov.au/Rubbish_and_Recycling/Compost_and_Worm_Farms/index.aspx Moonee Valley: www.mvcc.vic.gov.au.

Pipatyi R. and Savolainen I. "Role of Energy Production in the Control of Greenhouse Gas Emissions from Waste Management" VTT Energy (Technical Research Centre of Finland) 1997.

Sustainability Victoria 2010 "Towards Zero Waste Strategy 2009 –10 Progress Report" viewed 13.10.11 http://www.sustainability.vic.gov.au/ resources/ documents/TZW_Report 09_10.pdf.

In: Carbon Dioxide Emissions: New Research
Editors: M. Carpenter and E. J. Shelton
ISBN: 978-1-62257-436-0

Chapter 5

EVALUATION OF THE POTENTIAL CARBON DIOXIDE EMISSIONS REDUCTIONS FROM THE USE OF CHP SYSTEMS IN COMMERCIAL BUILDINGS

Pedro J. Mago, Rogelio Luck
and Amanda D. Smith
Department of Mechanical Engineering,
Mississippi State University, MS, US

ABSTRACT

Combined heat and power (CHP) systems generate electricity near the site of use and capture the waste heat for space heating or hot water. CHP systems have the potential to reduce the CO_2 emissions resulting from power production when compared to a reference system where all the power is purchased from the electricity grid. However, the potential to reduce emissions with the use of CHP systems depends on the electrical and thermal requirements of the building. This chapter presents a model to evaluate the potential emission reductions resulting from the implementation of CHP systems. The model is based on two main parameters. The first parameter, R_e, is the percentage of the average hourly electricity demand of the building satisfied by the CHP system, and the second parameter, R_h, is the fraction of the thermal load that is satisfied by the CHP system. In order to demonstrate how to apply the model, four different type of buildings located in Chicago, Illinois are evaluated for their potential to reduce CO_2 emissions using a CHP system.

NOMENCLATURE

$CF_{PEC,e}$	Primary energy conversion factor for electricity
$CF_{PEC,f}$	Primary energy conversion factor for fuel
CHP	Combined heating and power
$Cost_{CHP}$	Operational cost when CHP system is operated

$cost_e$	Electricity cost
$cost_f$	Fuel cost
$Cost_{ref}$	Operational cost of the reference case
E_b	Electricity required by the building
E_{excess}	Excess electricity available for storage or export
E_{grid}	Electricity required from the grid
E_m	Electric energy registered at the meter when the CHP system is operated
E_{CHP}	Electric energy supplied by the PGU of the CHP system
E_{useful}	Electricity produced by the PGU that is used by the building
EF_e	Emission conversion factor for delivered electricity
EF_f	Emission conversion factor for delivered natural gas
CDE_{CHP}	Emissions with CHP system in use
CDE_{ref}	Emissions from the reference case
F_b	Fuel energy required to satisfy the thermal demand of the building
F_{boiler}	Fuel energy consumed by the boiler
F_m	Fuel energy registered at the meter when CHP system is operated
F_{CHP}	Fuel consumed by the PGU of the CHP system
PGU	Power generation unit
Q_{boiler}	Heat provided by the boiler
Q_h	Heat required by the building
Q_{rec}	Heat recovered from the PGU
Q_{req}	Heat required from the CHP system in order to satisfy $\bar{Q}_h$
Q_{useful}	Heat recovered from the CHP that is used by the building
R_{boiler}	Fraction of thermal load supplied by the boiler
R_e	Fraction of electric load supplied by the CHP system
R_{grid}	Fraction of electric load supplied by the grid
R_h	Fraction of thermal load supplied by the CHP system
ΔCDE	Difference between CDE_{ref} and CDE_{CHP}
η_{boiler}	Boiler efficiency
η_{CHP}	CHP system efficiency
$\eta_{electric}$	CHP electrical efficiency
η_{hc}	Heating system efficiency of the building
η_{hrs}	Heat recovery system efficiency
η_{pgu}	PGU thermal efficiency
η_{CHP}	CHP thermal efficiency
ξ	Factor incorporating energy losses before the heat recovery system

1. Introduction

CHP systems capture heat produced as a byproduct of power production and deliver it to a building for space heating or hot water. A typical power plant rejects heat to the atmosphere as a result of power production, and in many cases a CHP system can make better use of its fuel energy by providing both power and heat as outputs from the power generation unit (PGU). As a result, a CHP system operating at a commercial building site may result in less

CO_2 emissions when compared with electricity purchased from the grid and heat produced with a boiler, with possible emission savings as high as 50% [1]. The International Energy Agency demonstrated that strategic use of CHP can help to slow the global increase in CO_2 production [2].

A CHP system may be optimized to reduce emissions, reduce cost, reduce primary energy consumption, or some desired combination of one or more of these factors [3-11]. The amount of electrical power produced by the PGU, the efficiencies of the components of the system, and the particular design and operational strategy are all factors which influence the amount of CO_2 emissions caused by a CHP system [7, 12-16].

In practice, the simplest strategy is for a CHP system to be sized such that it provides a constant amount of electrical power, called the base load, and for additional electrical power that is needed to come from the grid [17, 18]. This strategy is easily analyzed and it allows the CHP system to operate at a higher efficiency than a system which operates at partial load [19], so the case studies in this chapter all consider CHP systems that are base loaded. The PGU in each case produces power from natural gas, which is a common fuel for CHP systems.

Figure 1 shows a schematic of the CHP system evaluated in this chapter. Both electricity and waste heat are produced by the PGU. That electricity, along with electricity purchased from the grid, is used to meet the electrical need of the building.

The waste heat, rather than being rejected to the atmosphere, is captured by a heat recovery system. This recovered heat, along with supplemental heat produced by a boiler if it is needed, is used to supply the building's heating system. The heating system meets the thermal needs of the building by providing space heating and hot water.

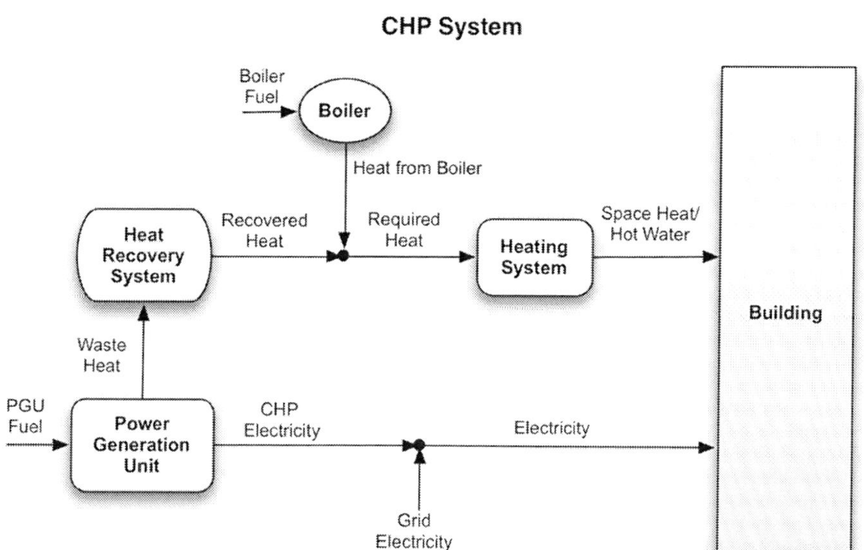

Figure 1. Schematic of a CHP System.

The CHP system is simulated for four different building types which each have different electrical and thermal needs, although the portion of the building's electricity that is provided by the CHP is designed to be constant as much as possible across the building types. The buildings are simulated in the same location in order to control for variations in electrical and thermal demand due to weather, and because the CO_2 emissions resulting from electrical power production vary by region and by utility [20]. Reasonable values are chosen for the various efficiencies associated with the CHP system and held constant from case to case.

2. Emissions from the CHP System Operation

To determine the potential to reduce emissions with the use of a CHP system, the emissions due to the CHP operation must be compared with the emissions from the reference building. The two scenarios for providing electricity and heat to the building are illustrated in Figure 2. The emissions caused by the reference system, shown on the left, result from the production and distribution of power plant electricity and the use of a boiler at the building site to supply heat. The emissions caused by the use of a base loaded CHP system, result from the CHP system operation as well as from supplemental electricity purchased from a power plant and supplemental heat provided by a boiler.

The variation of the CO_2 emissions with the use of the CHP system from the reference case can be determined as

$$\Delta CDE = CDE_{ref} - CDE_{CHP} \quad (1)$$

where CDE_{ref} are the CO_2 emissions from the reference case and CDE_{CHP} are the CO_2 emissions obtained when the CHP system is operated and can be calculated as

$$CDE_{ref} = E_b EF_e + F_b EF_f \quad (2)$$
$$CDE_{CHP} = E_m EF_e + F_m EF_f \quad (3)$$

where F_b is the fuel energy required to satisfy the thermal demand of the reference building, E_b is the electric demand of the reference building, E_m and F_m are the electricity consumption and the total fuel energy consumption registered at the electric meter and fuel meter, respectively, when the CHP system is operated, and EF_e and EF_f are the CO_2 emission conversion factors for delivered electricity and natural gas, respectively. The CO_2 emission conversion factors for delivered electricity strongly depend on the location where the facility is installed and on the fuel mix used to generate electricity in that particular location.

The electricity consumption and the total fuel energy consumption registered at the fuel meter and electric meter during the CHP operation can be determined as:

$$F_m = F_{CHP} + F_{boiler} \quad (4)$$
$$E_m = E_{grid} \quad (5)$$

where F_{CHP} is the CHP system fuel energy consumption, F_{boiler} is the boiler fuel energy, and E_{grid} is the electricity that needs to be imported from the grid.

The CHP system fuel energy consumption can be estimated as

$$F_{CHP} = E_{CHP}/\eta_{pgu} \quad (6)$$

where E_{CHP} is the electric energy that is supplied by PGU of the CHP system and η_{pgu} is the PGU thermal efficiency. Since in this chapter the PGU is assumed to operate at constant load, the efficiency of the PGU is assumed to be constant.

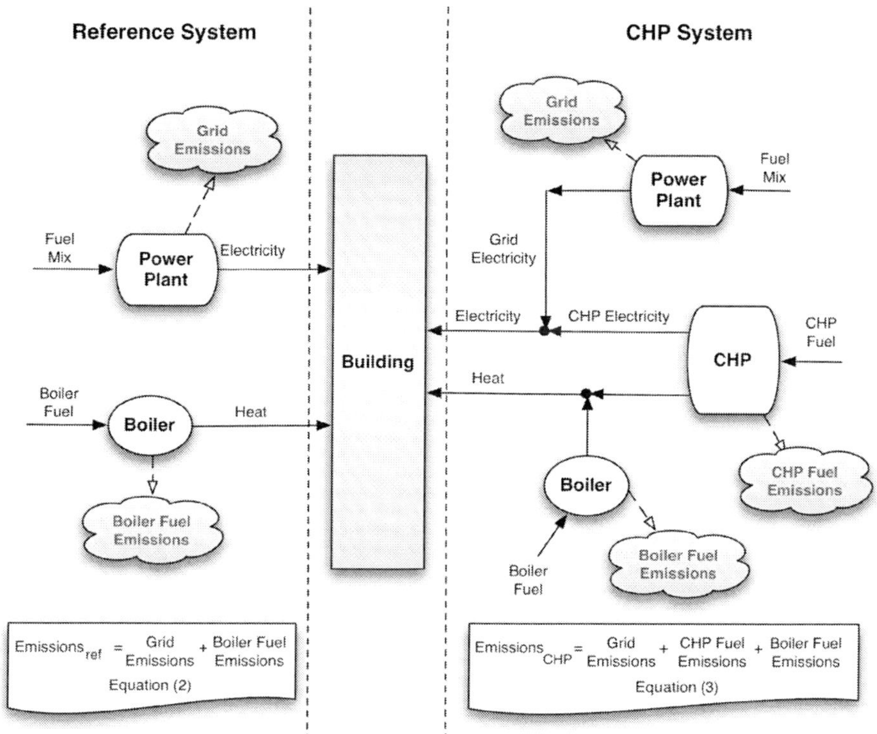

Figure 2. The two scenarios for providing electricity and heat to the building: Reference system and CHP system.

The electricity supplied by the PGU of the CHP system, i.e., the amount of base load, is assumed to be a fraction, R_e, of the hourly electricity needs of the building as follows:

$$E_{CHP} = R_e E_{b,ave} \tag{7}$$

where $E_{b,ave}$ represents the average hourly electricity needs of the building including electric equipment, lights, and the electricity needed for cooling.

To determine the boiler fuel energy, first, the heat recovered from the PGU needs to be calculated as:

$$Q_{rec} = \xi(F_{CHP} - E_{CHP})\eta_{hrs} \tag{8}$$

where η_{hrs} is the efficiency of the heat recovery system and ξ is a factor that accounts for PGU energy losses before the heat recovery system.

The heat required from the CHP system in order to meet the building's heating requirement Q_b is

$$Q_{req} = \frac{Q_b}{\eta_{hc}} \tag{9}$$

where η_{hc} is the efficiency of building heating system when the CHP system and/or boiler are used to meet the building's heating load.

If the heat recovered (Eq. 8) from the PGU is sufficient to satisfy the thermal load, a term called Q_{useful} can be used to represent only the portion of the recovered heat that is used to satisfy the thermal demand of the building, therefore,

$$Q_{useful} = \emptyset Q_{req} + (1 - \emptyset) Q_{rec} \tag{10}$$

where $\phi = 1$ if $Q_{rec} \geq Q_{req}$ or $\phi = 0$ if $Q_{rec} < Q_{req}$.

If the heat recovered is not sufficient to satisfy the building thermal load, the boiler supplies additional heat as follows:

$$Q'_{boiler} = Q_{req} - Q_{useful} \tag{11}$$

where Q'_{boiler} represents the additional heat needed from the boiler to meet Q_{req}, which is 0 when $Q_{rec} \geq Q_{req}$.

The total thermal load of the building can be expressed in terms of the heat supplied by the CHP system, Q_{CHP}, and the heat supplied by the boiler, Q'_{boiler}, as follows:

$$Q_b = Q_{CHP} + Q_{boiler} \tag{12}$$

where Q_{CHP} and Q_{boiler} can be determined as:

$$Q_{CHP} = \frac{Q_{useful}}{\eta_{hc}} \tag{13}$$

$$Q_{boiler} = \frac{Q'_{boiler}}{\eta_{hc}} \tag{14}$$

If the boiler provides heat to the building, the boiler fuel energy can be determined as

$$F_{boiler} = \frac{Q'_{boiler}}{\eta_{boiler}} \tag{15}$$

where η_{boiler} is the boiler efficiency.

Depending of the size of the PGU associated with the CHP system, the amount of generated electricity may be more or less than the electricity required by the building (E_b). Therefore, if

$$E_b > E_{CHP} \rightarrow E_{grid} = E_b - E_{CHP} \tag{16}$$

$$E_b < E_{CHP} \rightarrow E_{excess} = E_{CHP} - E_b \tag{17}$$

where E_{grid} is the amount of electricity required from the grid and E_{excess} is the amount of excess electricity that can be exported or stored for future use.

The contribution of the CHP system to satisfy the electric and thermal building loads can be defined as:

$$R_e + R_{grid} = 1 \rightarrow \frac{E_{CHP}}{E_b} + \frac{E_{grid}}{E_b} = 1 \tag{18}$$

$$R_h + R_{boiler} = 1 \rightarrow \frac{Q_{CHP}}{Q_b} + \frac{Q_{boiler}}{Q_b} = 1 \tag{19}$$

where R_e represents the fraction of the total electric load that is supplied by the CHP system, R_{grid} represents the fraction of the electric load that has to be imported from the grid, R_h is the fraction of the thermal load supplied by the CHP system and R_{boiler} represents the fraction of the thermal load supplied by the boiler.

The ratio of R_e to R_h can be expressed as

$$\frac{R_e}{R_h} = \frac{PHR_{CHP}}{PHR_b} \tag{20}$$

where PHR_{CHP} is the CHP power to heat ratio and PHR_b is the building power to heat ratio. Both of these parameters can be determined as:

$$PHR_b = \frac{E_b}{Q_b} \tag{21}$$

$$PHR_{CHP} = \frac{E_{CHP}}{Q_{CHP}} \tag{22}$$

If all the heat recovered is used by the building $(Q_{useful} = Q_{rec})$, using Eqs. (6), (8), and then Eq. (9) the ideal power to heat ratio ($PHR_{CHP,ideal}$) can be expressed as

$$PHR_{CHP,ideal} = \frac{\eta_{hc}\eta_{pgu}}{\xi\eta_{hrs}(1-\eta_{pgu})} \tag{23}$$

When the $PHR_{CHP} = PHR_{CHP,ideal}$ none of the heat produced by the CHP system is wasted, which is an ideal situation to maximize the benefits of the CHP system.

Therefore, for a fixed value of R_e, and a known value of PHR_b for a specific building, the $R_{h(ideal)}$ that corresponds to the $PHR_{CHP,ideal}$ for the CHP system can be calculated using Eq. (20)

$$R_{h(ideal)} = R_e \frac{PHR_b}{PHR_{CHP,ideal}} \tag{24}$$

The closer the value of the actual R_h to the $R_{h(ideal)}$ the better the performance of the CHP system. As with $PHR_{CHP,ideal}$, the $R_{h(ideal)}$ indicates that the heat produced by the CHP is useful to the building.

Therefore, a new ratio, *PR*, can be defined to compare the actual R_h to the $R_{h(ideal)}$ as follow:

$$PR = \frac{R_h}{R_{h(ideal)}} \tag{25}$$

In addition to the emissions reduction, the operational cost, the primary energy consumption and the overall CHP system efficiency and operational strategy are parameters that must be considered during the evaluation of the CHP system [3-5, 12, 17].

However, since this chapter is focused on CO_2 emissions, only the CHP system efficiency is used to compare the emissions obtained for the different evaluated scenarios.

The CHP system efficiency can be expressed as

$$\eta_{CHP} = \eta_{electric} + \eta_{thermal} \tag{26}$$

where $\eta_{electric}$ is the electric efficiency and $\eta_{thermal}$ is the thermal efficiency of the CHP system.

The electric efficiency can be determined as:

$$\eta_{electric} = \frac{E_{useful}}{F_{CHP}} \tag{27}$$

where E_{useful} is the portion of the electricity produced by the PGU that is used by the building.

In this chapter, the CHP system is base-loaded and sized to satisfy a fraction of the building's average hourly electricity needs, $E_{useful} = E_{pgu}$, therefore $\eta_{electric} = \eta_{pgu}$.

The thermal efficiency can be determined as:

$$\eta_{thermal} = \frac{Q_{useful}}{F_{pgu}} \tag{28}$$

3. RESULTS

To illustrate the model to estimate the potential emission reduction from the use of CHP systems presented in Section 2, four benchmark buildings developed by the Department of Energy were used to apply the model described in Section 2 [21].

The benchmark buildings were simulated over a year using Energy Plus [22] software using the weather data for Chicago, IL, and the output from the simulations, in the form of electric and thermal loads for the building, were used as inputs to the model presented in Section 2. Table 1 presents the different buildings selected, including the total floor area, and the results obtained from the simulations, as well as the PHR_b. The size of the selected buildings ranges from 511.2 m^2 (full service restaurant) to 4,180.8 m^2 (supermarket). Table 2 presents the CO_2 emission conversion factors for electricity and natural gas [23].

Initially, to compare the performance of the CHP system it was assumed that the ratio of electricity produced by the CHP to the total building electricity (R_e) was the same for all the evaluated buildings (0.2). Table 3 shows the size of the PGU used to simulate each building, which represents approximately 20% of the average electricity needed by the building in an hour.

Table 1. Building information and utility cost for the evaluated buildings in Chicago, IL [22]

	Full Service Restaurant	Outpatient	Supermarket	Small Hotel
Total Area (m^2)	511.2	3,804.0	4,180.8	4,013.6
Total Electricity (GJ)	1,126.2	4,880.2	5,929.2	2,104.5
Total Gas (GJ)	1,972.4	3,275.1	3,435.7	1,080.4
Thermal Load (GJ)	1577.9	4197.1	1667.9	2748.6
Building Power to heat ratio (PHR_b)	0.71	5.10	1.85	2.16

Table 2. CO_2 emission conversion factors conversion factors for electricity and natural gas for the city of Chicago [23]

	Electricity	Natural Gas
CO_2 Conversion Factor (g/MJ)	341.7	52.1

Table 3. PGU size used to simulate the evaluated buildings for R_e= 0.2

Building	PGU Size (kW)
Full Service Restaurant	7
Outpatient Health Care	31
Supermarket	38
Small Hotel	14

Table 4. CHP system parameters

Parameter	Value
PGU Efficiency, η_{pgu}	0.25
Factor that accounts for energy losses, ξ	0.95
Boiler efficiency, η_{boiler}	0.8
Heat recovery system efficiency, η_{hrs}	0.8
Building Heating System efficiency, η_{hc}	0.8

As mentioned before, the size of the PGU in relation to the building's electrical needs is held constant across all building types, which allows the influence of other parameters on the overall CHP system performance to be determined. Table 4 presents the CHP system parameters used in the results presented in this chapter.

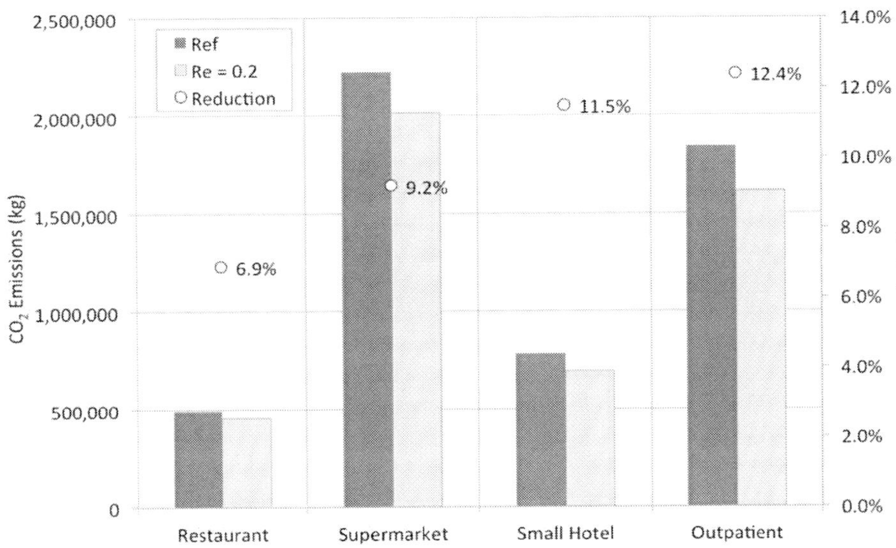

Figure 3. Comparison of the CO_2 emissions of the reference building with the CHP building.

Figure 3 illustrates a comparison of the CO_2emissions of the reference building and the building using a CHP system. In general it can be seen that the use of a CHP system reduces the CO_2emissions for all the evaluated buildings. The building that shows the highest percent reduction of CO_2emissions is the outpatient building, 12.4%. On the other hand, the building that shows the lowest percent reduction of CO_2 emissions is the restaurant,6.9%.

3.1. Effect of R_e on the CO_2 Emission Reduction

Figure 4 shows a comparison of the CO_2 emissions of the reference building and the building using a CHP system for different R_e values. The selected R_e values are 0.1, 0.2, and 0.3, representing 10% to 30% of the building's electrical needs being met by the CHP system. These values were chosen to be small in order to guarantee that no excess electricity is generated from the CHP system operation. Figure 4 illustrates that increasing R_e values reduced CO_2 emissions for all selected buildings. However, it is important to mention here, that even with 10% (R_e = 0.1) of the building electric load generated by the CHP system, significant reduction of CO_2 emissions is possible.

Figure 5 shows the reduction of CO_2 emissions for all the evaluated cases. As in Figure 3 the outpatient building shows the highest reduction of CO_2 emissions for R_e = 0.3 (18.5%) while the lowest reduction in CO_2 emissions for Re = 0.3 was obtained for the restaurant (12.7%). According to the results presented in Figures 4 and 5 it seems that increasing R_e is the best way to operate the CHP system in terms of emissions. However, it is important to consider that after certain values of R_e the CHP system will produce significant amounts of excess electricity for some cases and thermal energy will not be used in other cases, which reduces the overall CHP system efficiency. This will increase the CHP operational cost and may result in higher emission due to its lowered efficiency.

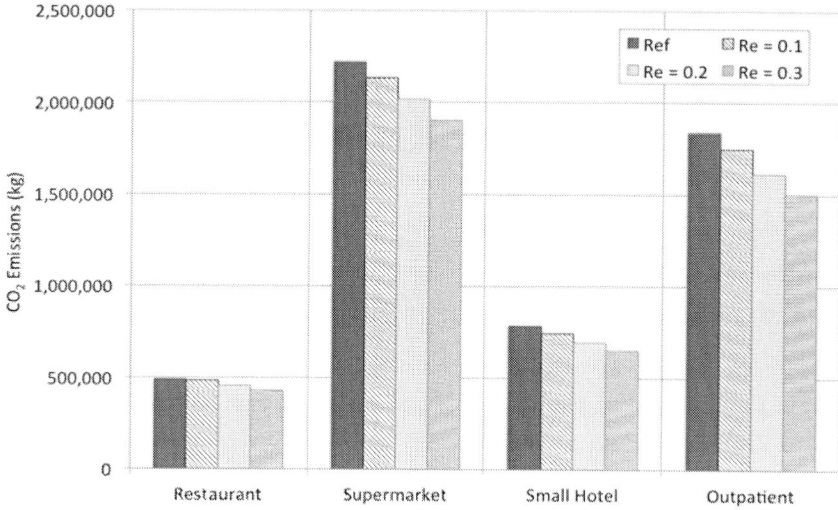

Figure 4. Carbon dioxide emissions of the reference building and the CHP-building for different R_e values.

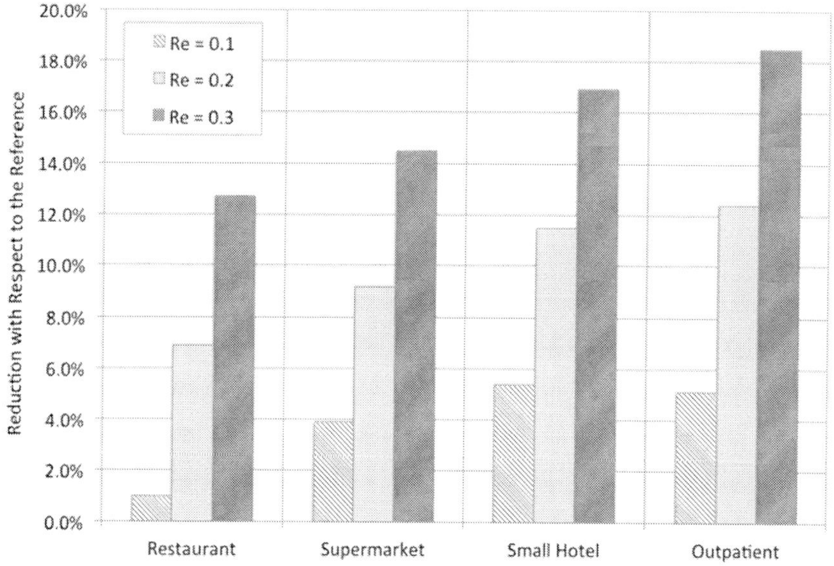

Figure 5. Carbon dioxide emissions Reduction of the CHP-building with respect o the Reference building for different R_e values.

3.2. Effect of R_h on the Emissions Reduction

Figure 6 illustrates the R_h values for all the evaluated buildings for different values of R_e. It can be seen that the value of R_h is different for all the evaluated buildings and that it increases with the increasing of R_e. Comparing the results presented in Figure 6 with Figure 5, it can be observed that the higher the R_h value the higher the reduction of emissions from the CHP system.

This can be explained since higher R_h values mean that the CHP system is providing most of the thermal load using the recovered heat which reduces the amount of fuel that otherwise would be necessary to satisfy the thermal demand of the building. Again, the overall CHP system efficiency is the only parameter that is used in this chapter in combination with the CO_2 emissions to analyze the evaluated cases.

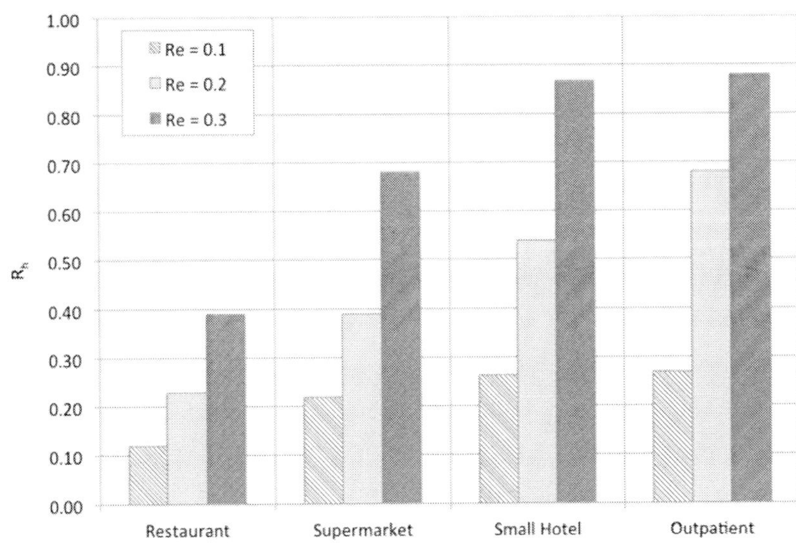

Figure 6. R_h values for the selected buildings for different values of R_e.

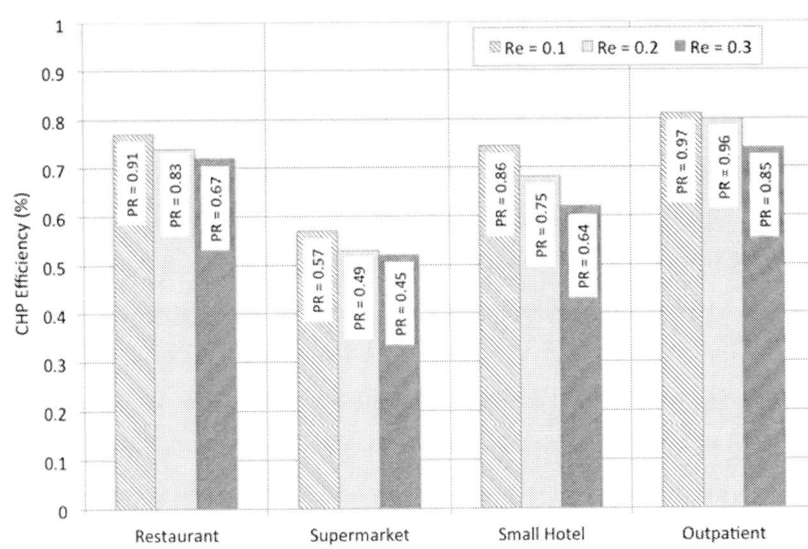

Figure 7. Efficiency of the CHP system and the associate PR values for the selected buildings for different values of R_e.

Figure 7 shows the efficiency of the CHP system and the associated *PR* values for the selected buildings for different values of R_e. This figure illustrates that increasing the R_e decreases the CHP efficiency. Because the CHP system generates more waste heat as a result of increased electricity production, some heat may not be used by the building, which decreases the overall CHP efficiency. In addition, it can be observed that the higher the *PR* values the better the CHP efficiency. This is also related to the amount of recovered heat not used by the facility. Higher *PR* values means that more of the recovered heat is useful to the building and therefore the CHP system is more efficient.

According to the results presented in this Chapter it seems that increasing R_e is the best way to operate the CHP system in terms of reducing the CO_2 emissions. However, it is important to consider that increasing R_e can also decrease the overall CHP efficiency.

CONCLUSION

This chapter presented a model to evaluate the potential emission reductions from the use of CHP systems. In order to demonstrate how to use the model, the model was applied to four different commercial buildings that were simulated in the city of Chicago. Results indicated that the use of a CHP system reduced the CO_2 emissions for the buildings considered. When the CHP system provided 20% of the electricity needed by the facility, the building that showed the highest reduction of CO_2 emissions was the outpatient building, 12.4%.

It was concluded that in general it is beneficial to have high R_h values, or to guarantee that the CHP provides a significant portion of the thermal demand of the facility, since this will result in better emissions from the CHP system operation. In addition, it is important to keep in mind that the R_h value should be selected to be close to the $R_{h(ideal)}$ since this guarantees higher CHP system efficiencies. The results presented in this chapter reflect the fact that the CO_2 emissions that could be obtained from the use of CHP system could be translated into economic savings with the right policies and incentives in place.

REFERENCES

[1] Meunier, F., 2002. "Co- and tri-generation contribution to climate change control," *Applied Thermal Engineering*, 22, pp. 703-718.

[2] International Energy Agency, 2008. "*Combined Heat and Power: Evaluating the Benefits of Greater Global Investment.*" Available at: http://www.iea.org/Papers/2008/chp_report.pdf.

[3] Mago, P.J., Chamra, L.M., and Hueffed, A. 2009. "A Review on Energy, Economical, and Environmental Benefits of the Use of CHP Systems for Small Commercial Buildings for the North American Climate." *International Journal of Energy Research*, 33, pp. 1252-1265.

[4] Smith, A.D., Fumo, N., and Mago, P.J., 2011. "Spark Spread – A Screening Parameter for Combined Heating and Power Systems," *Applied Energy*, 88, Issue 5, pp. 1494-1499.

[5] Smith, A.D., Mago, P.J., and Fumo, N., 2011. "Emissions Spark Spread and Primary Energy Spark Spread – Environmental and Energy Screening Parameters for Combined Heating and Power Systems," *Applied Energy,* 88, Issue 11, pp. 3891-3897.

[6] Fumo, N., Mago, P.J., and Chamra, L.M., 2009. "Energy and economic evaluation of cooling, heating, and power systems based on primary energy," *Applied Thermal Engineering*, 29, pp. 2265-2671.

[7] Mago, P.J., Fumo, N., and Chamra, L.M., 2009. "Performance Analysis of CCHP and CHP Systems Operating Following the Thermal and Electric Load." *International Journal of Energy Research*, 33, pp. 852–864.

[8] Ehyaei, M.A., and Mozafari, A. 2010. "Energy, economic and environmental (3E) analysis of a micro gas turbine employed for on-site combined heat and power production," *Energy and Buildings*, 42, pp. 259-264.

[9] Kavvadias, K.C. and Maroulis, Z.B. 2010. "Multi-objective optimization of a trigeneration plant," *Energy Policy*, 38, pp. 945-954.

[10] Jiang-Jiang, W., Chun-Fa, Z., and You-Yin, J. 2010. "Multi-criteria analysis of combined cooling, heating, and power systems in different climate zones in China." *Applied Energy*, 87, pp. 1247-1259.

[11] Piacentino, A. and Cardona, F. 2008. "An original multi-objective criterion for the design of small-scale polygeneration systems based on realistic operating conditions," *Applied Thermal Engineering*, 28, pp. 2391-2404.

[12] Hueffed, A.K., and Mago, P.J., 2010. "Influence of prime mover size and operational strategy on the performance of combined cooling, heating, and power systems under different cost structures," *Journal of Power and Energy,* 224(5), pp. 591-605.

[13] Dorer, V., and Weber, A., 2009. "Energy and carbon emission footprint of micro-CHP systems in residential buildings of different energy demand levels," *Journal of Building Performance Simulation*, 2(1), pp. 31-46.

[14] Sayane, S. and Shokrollahi, S. "Selection and Sizing of Prime Movers in Combined Heat and Power Systems." *Proceedings of ASME Turbo Expo,* 2004, Vienna, Austria, pp. 613-621.

[15] Zogg, R., Roth, K., and Brodrick, J. 2005. "Using CHP Systems in Commercial Buildings." *ASHRAE Journal*, September, pp. 33-36.

[16] Al-Sulaiman, F. A., Hamdullahpur, F. and Dincer, I. 2011. Trigeneration: A comprehensive review based on prime movers. *International Journal of Energy Research*, 35 (3), pp. 233-258.

[17] Fumo, N., Mago, P.J., and Smith, A.D., "Analysis of Combined Cooling, Heating, and Power Systems Operating Following the Electric Load and Following the Thermal Load Strategies with No Electricity Exporting," *Journal of Power and Energy*, 225(8), pp. 1016-1025.

[18] Mago, P.J., and Luck, R. 2011. "Evaluation of a Base-Loaded Combined Heating and Power System with Thermal Storage for Different Small Buildings Applications." *International Journal of Energy Research,* doi: 10.1002/er.1892.

[19] Sclafani, A., and Beyene, A., 2008. "Sizing CCHP Systems for Variable and Non-Coincident Loads: Part 1 – Load Profiling and Equipment Selection," *Cogeneration and Distributed Generation Journal*, 23(3), pp. 6-19.

[20] The US Department of Energy, Environmental Protection Agency, *"Power Profiler."* Available at: http://www.epa.gov/cleanenergy/ energy-and-you/how-clean.html.

[21] Torcellini, P., Deru, M., Griffith, B., and Benne, K., 2008. "DOE Commercial Building Benchmark Models," *ACEEE Summer Study on Energy Efficiency in Buildings, Pacific Grove*, CA, August 17-22, 2008.

[22] U.S. Department of Energy, 2012. "*Energy Plus Energy Simulation Software,*" Available at: http://apps1.eere.energy.gov/buildings/ energyplus/.

[23] U.S. Department of Energy, May 2009. "New construction bench- mark data files," Available from http://www1.eere.energy.gov/buildings/commercial_initiative/new_ construction.html.

In: Carbon Dioxide Emissions: New Research
Editors: M. Carpenter and E. J. Shelton

ISBN: 978-1-62257-436-0

Chapter 6

BUILDING INTEGRATED RENEWABLE ENERGY TECHNOLOGIES: EMBODIED ENERGY, ECONOMIC ANALYSIS AND POTENTIAL OF CO_2 EMISSION MITIGATION

Arvind Chel[*]
Centre for Energy Studies,
Indian Institute of Technology (IIT) Delhi, New Delhi, India

ABSTRACT

Owing to global energy crisis, energy conservation through energy efficient buildings has acquired prime importance. Hence, this chapter deals with various renewable energy technologies that can be integrated with building. The three main aspects which should be given consideration for energy saving in a building include first and foremost the building design, secondly the low energy building materials in building construction and lastly the use of renewable energy technologies for various applications. The solar passive heating/cooling design features like sunspace, Trombe walls, earth to air heat exchangers, adobe house, day-light etc can lead to considerable energy saving at the stage of building design. Further, low embodied energy building materials should be chosen like fly ash bricks; fiber reinforced bricks; and stabilized mud blocks etc. which reduce energy consumption at building construction stage. The use of such low embodied energy materials has gained importance nowadays. The energy requirements of building can be cut down to minimum by use of renewable energy technologies like solar photovoltaic, wind turbines, solar thermal devices etc. These technologies are discussed in this chapter with their economic analysis and environmental impacts. In the present alarming situation of global warming, it is known that about 1 kg of CO_2 gets emitted in to the environment corresponding to generation of 1 kWh unit of electrical energy from the coal thermal power plants. Hence, for sustainable future of the world, there is urgent need for use of renewable energy sources to mitigate CO_2 emissions. Today, alternative renewable energy technologies are becoming prime important to meet at least 10 % of the total building energy requirements in European countries like U.K.

[*] Corresponding author's e-mail: dr.arvindchel@gmail.com, Tel.: +91-9968144689 (India)

1. Introduction

The dwelling and habitat are the most proximate and visible symbols of present day lifestyles. About 30-40 % of total global basic resources are invested in building sector, which is growing at a rapid pace. The common aspirants are invariably linked to making buildings as comfortable and convenient as possible. It is in this pursuit that the present day buildings have become the third largest consumer of fossil energy after industry and agriculture. The Asia-Link Programme is an initiative by the European Commission to promote and spread the knowledge on Sustainable Built Environment quickly and widely for people at all levels in general, all the reading materials can be viewed and downloaded from website (www. ChinaSBE.com). Energy consumed in the building sector constitutes a significant proportion of total energy consumption. Sources place the amount of energy expended in the building sector in Europe to about 40–45% of total energy consumption [1]; about two thirds of this amount is used in private buildings. Other sources claim, that in industrialized countries, energy usage in buildings is responsible for approximately 50% of carbon dioxide emissions [2-4]. Hence there exists a tremendous potential to conserve energy in buildings. Energy conservation measures are developed for newly constructed buildings and for buildings under refurbishment. However, to achieve a significant reduction in energy consumption apart from the standard energy-efficiency methods, innovative technologies should be implemented, including renewable energy [5, 6].

2. Energy Conservation in Building

There are three broad ways to reduce the energy consumption of building to mitigate CO_2 emissions. These are described as follows:

a. Passive building design
b. Low embodied energy construction materials and
c. Building integrated with renewable energy technologies

2.1. Passive Building Design

The most sustainable energy technique is to conserve energy as much as possible. Passive solar building design can aid energy conservation efforts because building design is directly related to energy use. Buildings with passive solar building designs naturally use the sun's energy for free of charge heating, cooling and day-lighting.

This reduces the need to consume energy from other sources and provides a comfortable environment inside. The principles of passive solar design are compatible with diverse architectural styles and building techniques [7].

2.1.1. Passive Solar Design Principles

Passive solar design integrates a combination of building features to reduce or even eliminate the need for mechanical cooling and heating and daytime artificial lighting [8].

Designers and builders pay particular attention to the sun to minimize heating and cooling needs. The design does not need to be complex, but it should involve knowledge of solar geometry, window technology, and local climate. Given the proper building site, virtually any type of architecture can integrate passive solar design [8].

The basic natural processes that are used in passive solar energy are the thermal energy flows associated with radiation, conduction, and natural convection. When sunlight strikes a building, the building materials can reflect, transmit, or absorb the solar radiation.

Additionally, the heat produced by the sun causes air movement that can be predictable in designed spaces. These basic responses to solar heat lead to design elements, material choices and placements that can provide heating and cooling effects in a home. Passive solar energy means that mechanical means are not employed to utilize solar energy. There are some rules of thumb which must be considered for effective solar energy utilization through passive solar systems [9]. The building and safety regulation issues are also very important while designing the solar passive building [9].

2.1.2. Passive Solar Heating

The goal of all passive solar heating systems is to capture the sun's heat within the building's elements and release that heat during periods when the sun is not shining.

At the same time that the building's elements (or materials) is absorbing heat for later use, solar heat is available for keeping the space comfortable (not overheated). Two primary elements of passive solar heating required are as follows:

- South facing glass for northern region and vice versa
- Thermal mass to absorb, store, and distribute heat

There are three approaches to passive solar heating systems - direct gain, indirect gain, and isolated gain.

I. *Direct gain* - In this system, the actual living space is a solar collector, heat absorber and distribution system. South facing glass admits solar energy into the house where it strikes directly and indirectly thermal mass materials in the house such as masonry floors and walls as shown in Figure 1. The direct gain system will utilize 60-75% of the sun's energy striking the windows. In a direct gain system, the thermal mass floors and walls are functional parts of the house. The thermal mass absorbs solar radiations during daytime and radiates the heat energy during night time into the living space [10] as shown in Figure 1.
II. *Indirect gain* - In an indirect gain of solar passive heating system, thermal mass is located between the sun and the living space. The thermal mass absorbs the sunlight that strikes it and transfers it to the living space by conduction. The indirect gain system will utilize 30-45% of the sun's energy striking the glass adjoining the thermal mass. There are three types of indirect gain systems

- Thermal storage wall systems (or Trombe wall)
- Water wall
- Roof pond systems

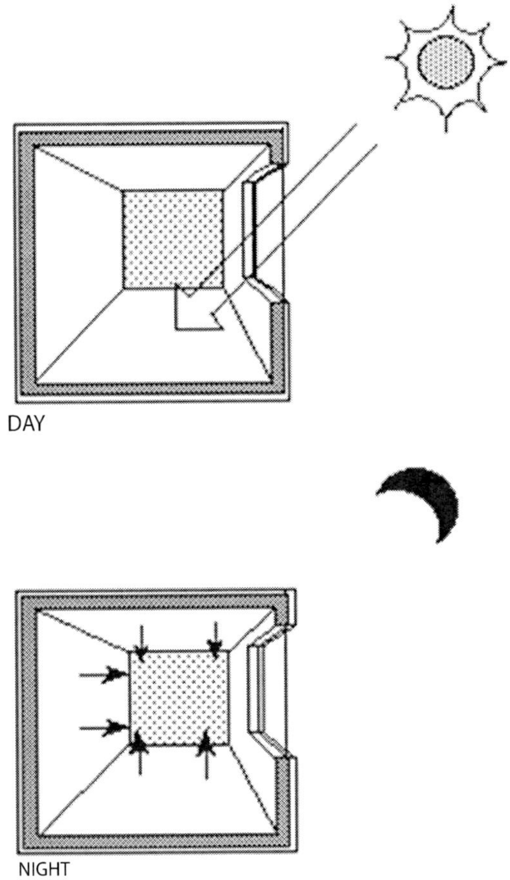

Figure 1. Direct gain: Thermal mass absorbs heat in day and radiates in night [10].

In indirect gain solar passive heating system, Trombe wall absorbs and stores heat during the day. Excess heat is carried out by passage air between wall and glass through thermosyphon principle in to the interior space as shown in Figure 2. At night Trombe wall vents are closed and the storage wall radiates heat into the interior space as shown in Figure 2.

In water wall indirect solar passive heating system, the wall is composed of water stored in the transparent/opaque containers. During daytime water absorbs solar heat and radiates heat during night as shown in Figure 3 [11].

An indirect gain solar passive design which provides both heating and cooling is the thermal pond approach, which uses water encased in ultraviolet ray inhibiting plastic beds underlined with a dark color, that are placed on a roof. Hence, this system is known as roof pond solar passive heating/cooling system. In warm and temperate climates with low precipitation, the flat roof structure serves directly as a ceiling for the living spaces below (Figure 4) thereby facilitating heating and cooling for the spaces below. In colder climates, where heating is more desirable, attic ponds under pitched roof glazing are effective. Winter heating occurs when sunlight heats the water, which then radiates energy into the living space as well as absorbs heat within the water thermal mass for nighttime distribution.

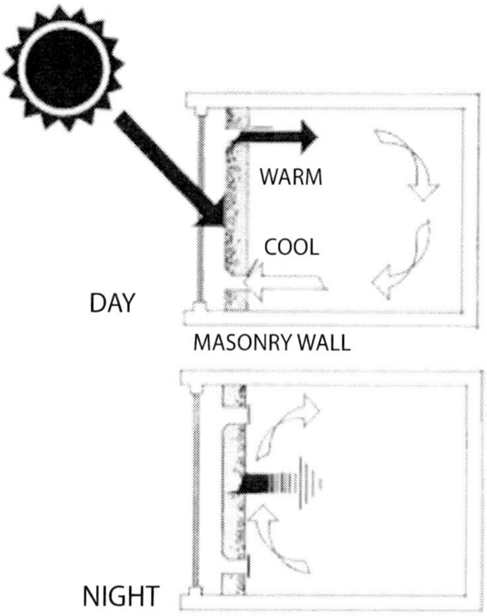

Figure 2. Thermal mass wall (or Trombe wall) operation in day and night [11].

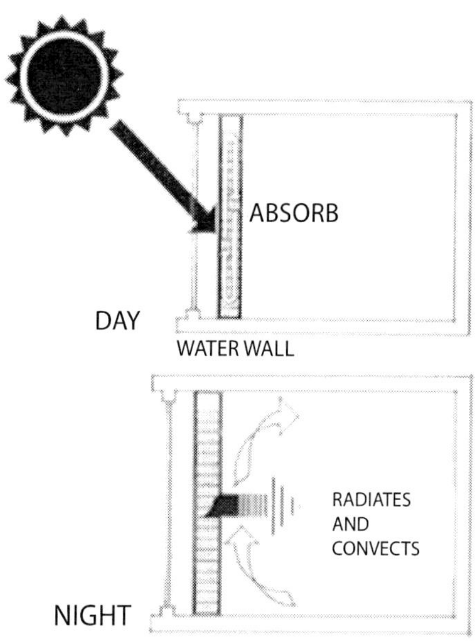

Figure 3. Water wall absorbs and stores heat during the day and radiates in night [11].

During nighttime in winter months, movable insulation is covered over the roof pond and hence it radiates heat in the interior living space as shown in Figure 4. During daytime in summer months, the movable insulation is covered over the roof pond so that solar heat gain is minimized and water absorbs heat from the room to provide the cooling effect inside the living space. While during nighttime in summer months, the movable insulation is removed and the water radiates heat outside the room by absorbing the heat from the interior living space [11]. One of the major advantages of this approach is that it allows all rooms to have their own radiant energy source with little concern about the orientation of the structure or optimal building form [11].

iii. *Isolated gain* - An isolated passive solar heat gain system has its integral parts separate from the main living area of a house [10]. Isolated gain involves utilizing solar energy to passively move heat from or to the living space using a fluid, such as water or air by natural convection or forced convection [12]. Examples are sunroom and convective loop through flat plate air collector to a storage system in the house [10]. The ability to isolate the system from the primary living areas is the point of distinction for this type of system. The isolated gain system will utilize 15-30% of the sunlight striking the glazing toward heating the adjoining living areas. Solar energy is also retained in the sunroom itself.

iv. *Sunroom* (or solar greenhouse): It integrates a combination of direct gain and indirect gain system features. Sunlight entering the sunroom is retained in the thermal mass and air of the room. Sunlight is brought into the house by means of conduction through a shared mass wall in the rear of the sunroom, or by vents that permit the air between the sunroom and living space to be exchanged by convection [10]

Flat plate solar collector: It is attached to the structure separately and uses a fluid (liquid or air) to collect solar heat gain. The heat is transferred through ducts or pipes by natural convection to a storage area - comprised of a bin (for air) or a tank (for liquid), where the collected cooler air or water is displaced and forced back to the collector. If air is used as the transfer medium in a convection loop, heated air coming from the collector is usually directed into a rock (or other masonry mass material) bin where heat is absorbed by the rocks from the air. As the air passes its heat to the rocks it cools, falls to the bottom of the bin and is returned to the collector completing the cycle. At night the interior space of the structure is heated by convection of the collected radiant energy from the rock bin. If water is the transfer medium, the process works in much the same way except that heat is stored in a tank, and as hot water is introduced, cooler water is circulated to the collector. In naturally occurring convection systems (non-mechanically assisted) collectors must be lower than storage units, which must be lower than the spaces to be heated [11]

2.1.3. Passive Solar Cooling

A combination of proper insulation, energy-efficient windows and doors, daylighting, shading, and ventilation will usually keep homes cool with a low amount of energy use [13]. The approaches include use of operable windows, wing walls and thermal chimney.

Natural ventilation can be created by providing vents in the upper level of a building to allow warm air to rise by convection and escape to the outside. At the same time cooler air can be drawn in through vents at the lower level. This lower vent is provided where there are trees planted besides the building to provide shade for cooler outside air [14].

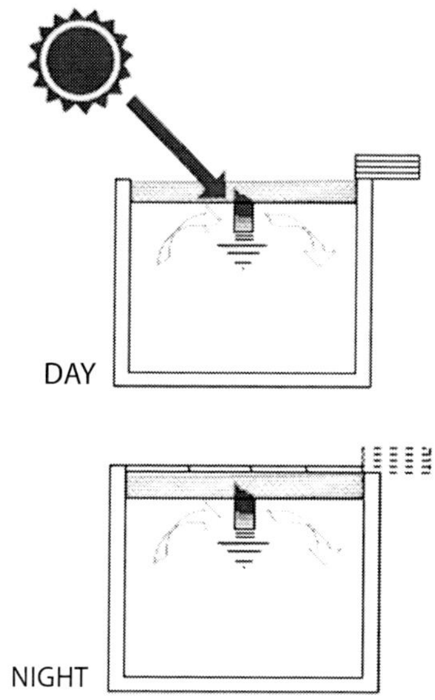

Figure 4. Roof pond stores heat daytime and radiates in nighttime for winter [11].

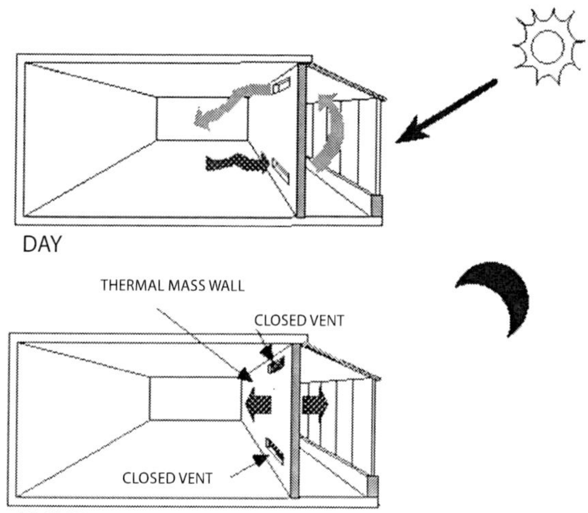

Figure 5. Day and night operation of a Sunroom isolated heat gain system [10].

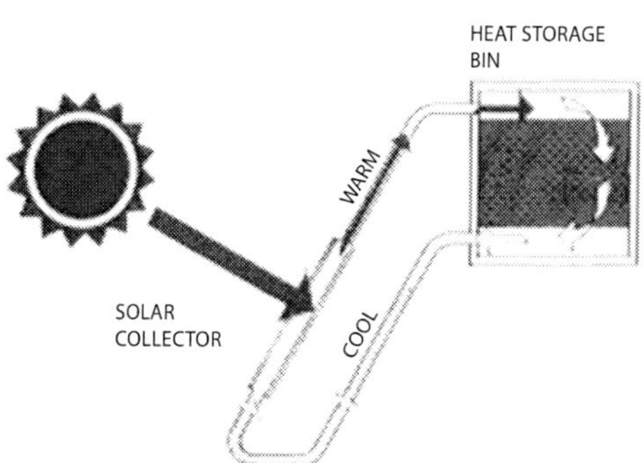

Figure 6. Flat plate solar heat gain collector with water/air convection loop [11].

I) Ventilation and Operable Windows

- Place operable windows on the south exposure.
- Casement windows offer the best airflow. Awning (or hopper) windows should be fully opened or air will be directed to ceiling. Awning windows offer the best rain protection and perform better than double hung windows.
- If a room can have windows on only one side, use two widely spaced windows instead of one window [15].

II) Wing Walls

Wing walls are vertical solid panels placed alongside of windows perpendicular to the wall on the windward side of the house.

Wing walls will accelerate the natural wind speed due to pressure differences created by the wing wall [15].

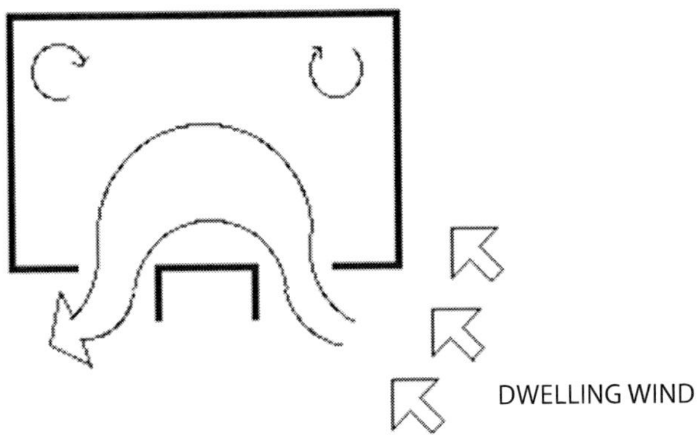

Figure 7. Top view of wing walls airflow pattern [15].

III) Thermal Chimney

A thermal chimney employs convective currents to draw air out of a building. By creating a warm or hot zone with an exterior exhaust outlet, air can be drawn into the house ventilating the structure. Thermal chimneys can be constructed in a narrow configuration (like a chimney) with an easily heated black metal absorber on the inside behind a glazed front that can reach high temperatures and be insulated from the house. The chimney must terminate above the roof level. A rotating metal scoop at the top which opens opposite the wind will allow heated air to exhaust without being overcome by the prevailing wind. Thermal chimney effects can be integrated into the house with open stairwells and atria as shown in Figures 8 and 9. This approach can add into the aesthetic of the home.

IV) Other Ventilation Strategies

- Make the outlet openings slightly larger than the inlet openings.
- Place the inlets at low to medium heights to provide airflow at occupant levels in the room.

Passive solar building cooling design is used to (1) slow the rate of heat transfer into a building in the summer, and (2) remove unwanted heat from a building. The principles of physics are holistically integrated into the exterior envelope. This is much easier to do in new construction. It involves a good understanding of the mechanisms of heat transfer: heat conduction, convective heat transfer, and thermal radiation (primarily from the sun) [16].

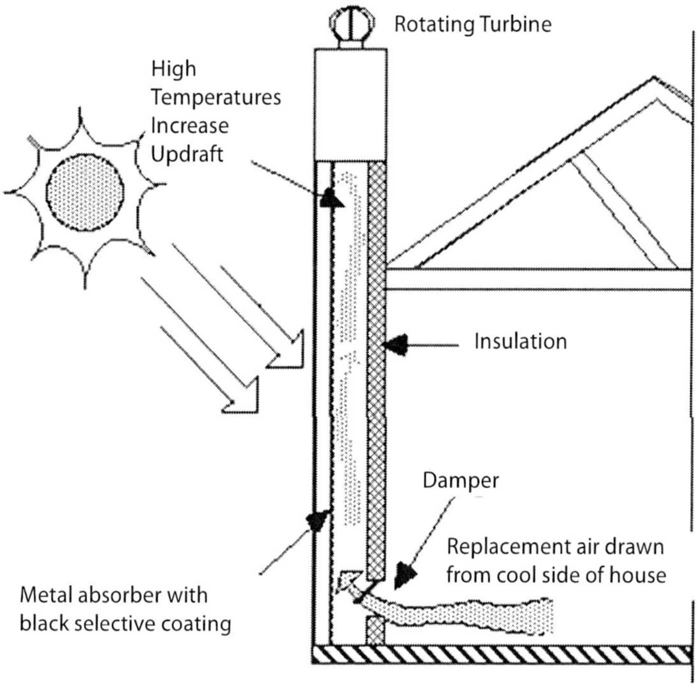

Figure 8. Thermal chimney [15].

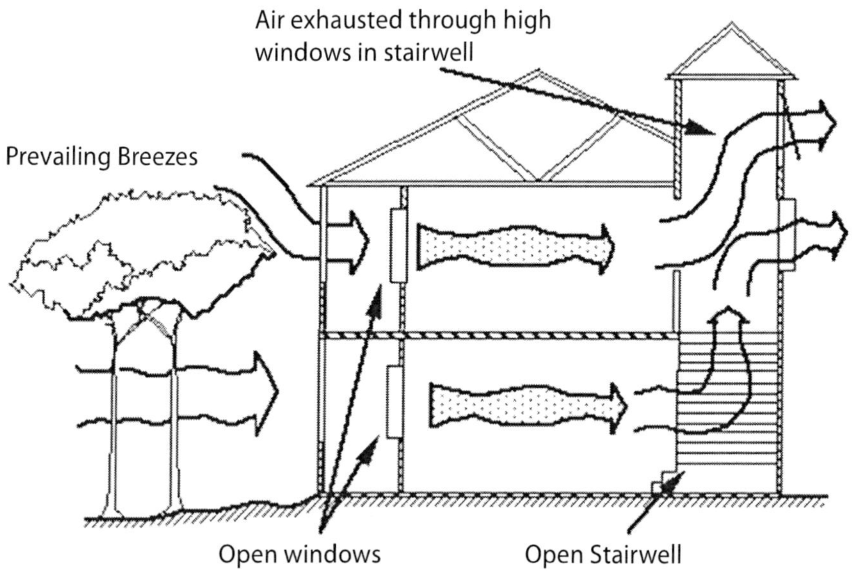

Figure 9. Thermal chimney effect built into home [15].

2.2. Embodied Energy in Buildings

An important goal for the building sector is to produce buildings with minimum environmental impacts. Energy use is a central issue as enormous energy is used in the construction of buildings [17]. Hence it is important to consider low embodied energy materials that reduce the energy in the building construction. Embodied energy is the energy consumed by all of the processes associated with the production of a building, from the acquisition of natural resources to product delivery, including mining, manufacturing of materials and equipment, transport and administrative functions [18].

Energy in buildings can be categorized into two types: (1) energy for the maintenance/ servicing of a building during its useful life, and (2) energy capital that goes into production of a building (embodied energy) using various building materials.

Study of both the types of energy consumption is required for complete understanding of building energy needs [19].

Presently the embodied energy of building materials contributes anywhere from 15 to 20% of the energy used by a building over a 50 year period. Homeowners have tremendous influence as to what material are used and can specify those materials with low embodied energy, thus reducing the amount of fossil-fuel energy used during production [20]. Use of low embodied energy materials in buildings can greatly reduce the energy consumption in buildings and also minimize the environmental impacts of building construction. Table 1 gives the embodied energy of the commonly used construction materials.

The construction industry is one of the largest in terms of employing manpower and volume of materials produced (cement, brick, steel and other materials).

Demand and supply gap for residential buildings is increasing every year. Cement (>75 million tones per annum), steel (>10 million tonnes per annum) and bricks (>70 billion per annum) are the largest and bulk consumption items in the Indian construction industry. Minimizing the consumption of the conventional materials by using alternative materials, methods and techniques can result in scope for considerable energy savings as well as reduction of CO_2 emission [19].

The modern tendency is to use energy intensive material in buildings like Aluminium, steel and glass. Hence, the consumption of these metals should be kept to a minimum, in order to keep the energy in a building low. Varieties of materials are used for the construction of masonry walls.

It is important to look for alternative materials with low embodied energy. Five types of building blocks viz. stone, burnt clay brick, soil–cement block, hollow concrete block and steam cured mud block were considered for embodied energy analysis and it was found that stone blocks do not consume any thermal energy, whereas burnt clay bricks consume maximum amount of energy among the alternatives studied. Soil–cement block with 6% cement is the most energy efficient block consuming only 23.5% of burnt clay brick energy. Soil–cement block and hollow concrete block with 7–8% cement have similar embodied energy values, i.e. about 30% of the burnt clay brick energy. Steam cured mud blocks consume about 60% of burnt clay brick energy. This can be attributed to high percentage of lime and fuel used for steaming operations. Building materials are transported over distances in excess of 100 km in many urban centers in India. Diesel energy spent for transportation could be about 5–10% of energy spent during manufacturing process for burnt clay bricks.

Energy spent in transporting high-energy materials like steel and cement is negligible when compared to the energy spent in the manufacture of these materials.

Table 1. Embodied primary energy of building materials [21]

Ranking	Building Material	Primary Energy Input (MJ/kg)
Very high energy	Aluminium	200-250
	Plastics	50-100
	Copper	100+
	Stainless steel	100
High energy	Steel	20-60
	Lead, Zinc	25+
	Glass	12-25
	Cement	5-8
	Plaster board	8-10
Medium energy	Lime	3-5
	Clay bricks and tiles	2-7
	Gypsum plaster	1-4
	Concrete	
	In situ	0.8-1.5
	Blocks	0.8-3.5
	Precast	1.5-8
	Sand-lime brick	0.8-1.2
	Timber (sawn)	0.1-5
Low energy	Sand, aggregate	<0.5
	Flyash, RHA, volcanic ash	<0.5
	Soil	<0.5
	Adobe	<0.2

The embodied energy analysis of various buildings components reveals that the energy content of burnt clay brick masonry is 2141 MJ/m^3. The soil–cement block masonry is most energy efficient at one-third the energy of burnt clay brick masonry. While the concrete block masonry has about 40–45% of energy content of burnt clay brick masonry. The use of stabilized mud blocks (SMB) filler blocks in solid reinforced cement (RC) roof/floor slabs leads to 20% reduction in energy content. Masonry vault roofs are more energy efficient than solid RC slab. Tile roofs have least energy content when compared with other roofing systems. A comparison of three types of buildings a multi-storey building; conventional two-storey load bearing brickwork building and another two-storey building using alternative building technologies has been done. It is found that the embodied energy of multi-storey RC framed structure building is the highest at 421 GJ (21 tonnes of coal equivalent) for 100 m^2 built-up area. Building with load bearing masonry structure using burnt clay bricks and RC slab has 30% less embodied energy when compared to RC framed structure building. The use of energy efficient alternative building technologies can result in considerable reduction in the embodied energy of the buildings. Load bearing soil–cement block masonry and SMB filler slab has resulted in 62% reduction in embodied energy when compared to RC framed structure building and 45% reduction when compared with burnt clay brick masonry and RC solid slab building [19].

The results of the above discussed work give useful tips for selecting an energy efficient building technology leading to considerable reduction in embodied energy of the building as a whole. Even though the results pertain to Indian conditions, many other developing nations have similar construction practices, where these results can be conveniently extrapolated and used.

Adobe (or mud) houses that use adobe, cow dung and other very low energy intensive material lead to very low emission of CO_2 in comparison to concrete or brick buildings. Based on embodied energy analysis, the energy payback time for the mud-house was determined as 18 years [22]. The embodied energy per unit floor area of R.C.C. building (3702.3 MJ/m^2) is quiet high as compared to the mud-house (2298.8 MJ/m^2) [22]. The annual heating and cooling energy saving potential of the mud-house was determined as 1481 kWh/year and 1813 kWh/year respectively for New Delhi composite climate [22]. The total mitigation of CO_2 emissions due to both heating and cooling energy saving potential was determined as 5.2 metric tons/year [22]. The annual carbon credit potential of mud-house was determined as € 52/year. Similar results were obtained for the different climatic locations in India. The mitigation of CO_2 emissions was determined as 58 metric tons due to construction of renovated mud-house of floor area about 94 m^2 compared to RCC structure house [22]. The carbon credits earned was determined as € 580 due to mitigation of CO_2 emissions from construction of mud-house instead of R.C.C. structure building (assuming € 10/metric Ton of CO_2 mitigation) [22]. Hence, total embodied energy of building can be substantially reduced when energy efficient/alternative low embodied energy building materials are used.

2.3. Building Integrated Renewable Energy Technologies

Renewable energy is derived from natural processes that are continuously replenished. The various forms of renewable energy derive directly from the sun, or from heat generated deep within the earth. The electricity and heat generated from solar, wind, ocean, hydropower, biomass, geothermal resources, and biofuels and hydrogen derived from renewable resources are also included in this definition [23]. The renewable energy technologies include solar power, wind power, hydroelectricity, micro-hydro, biomass and biofuels [24].

According to global status report 2007, about 18% of global final energy consumption in 2006 came from renewables, with 13% coming from traditional biomass, such as wood-burning. Hydropower was the next largest renewable source providing 3%, followed by hot water/heating which contributed 1.3%. Modern technologies, such as geothermal, wind, solar, and ocean energy together provided some 0.8% of final energy consumption [25]. Therefore the technical potential for their use is enormous, which exceeds all other readily available sources [26].

Significant amount of heat and electricity needs of buildings can be effectively covered by using solar thermal collectors and photovoltaic. In the coming years, other renewable energy sources (RES) such as wind turbines; biomass and hydrogen (produced only from RES) can be also applied, minimizing the use of the conventional energy sources. RES and nuclear energy can be considered the alternative energy sources to avoid greenhouse effect.

Between these two energy sources, only RES are clean and compatible with the environment, are almost uniformly distributed globally, can be easily used from all people with minimum market trust and ownership undertaking and of course are inexhaustible [27].

Renewable energy technologies have several benefits such as sustainability and security of energy supply, increased employment, long lifetime of energy systems etc. Even if the cost of solar energy systems remains in high levels it seems to be in compliance with European and international commitments. This is due to the fact that solar technology is a very friendly one in environmental terms for buildings and urban applications. This kind of technology is also highly important for the economy of most countries as they can replace the expensive and imported conventional energy sources (oil, gas, coal and nuclear fuels). Solar Energy Systems can be applied in a very harmonic way on buildings to cover the heating, cooling, electricity and lighting needs. The facades and the horizontal or inclined roofs of houses, hotels, athletic centers, etc, constitute appropriate surfaces for an expanded use of solar thermal collectors and photovoltaic panels.

Buildings can be designed according to the bioclimatic architecture for the minimization of the energy needs and the environmental impact of them, using new heat-insulating materials and special glasses (e.g. smart windows), which reduce effectively thermal losses during the winter and energy consumption for cooling during the summer. Under these aspects, the prospective energy savings in the buildings (especially in the new buildings) can be more than 50% of the energy consumption of standard buildings and become a regular procedure for the built environment construction [27].

The installation of devices and active solar energy units is related with their cost increase and their harmonization with building's architecture and the environment. Solar energy systems are also preferred for aesthetic reasons, in order to avoid the negative phenomena of diesel engines for heat and electricity (e.g. smokes, chimneys, etc). It is also important to apply them if they are harmoniously implemented into the existing, local, natural particularities of the environment through good planning and wise environmental studies.

The exploitation of solar energy systems towards sustainable development applications could take the form of the creation of innovative buildings, equipped with bioclimatic features aiming at the saving of energy. As it is known that the sector of buildings is responsible for about 35% of the final energy consumption and for the 40% of the gas emissions, it is estimated that the saving of energy can arrive at 60% when solar energy systems are used for heating and cooling purposes. Besides, a new directive from European Commission (EC) has been placed into force with regard to the obligation of energy saving in the newly constructed buildings. Thus, RES application to buildings with improved performance and aesthetic integration can result to the rising of the standards of living [27]. There are following building integrated solar energy systems adopted worldwide:

I) Flat Plate Thermosyphon Units (FPTU) and Integrated Collector Storage (ICS)

These systems are small size solar water heaters, aiming to cover domestic needs of about 100–200 liters of hot water per day [27,28]. ICS solar systems are simpler and have lower cost than FPTU systems, as they consist of solar collector and water storage tank mounted together in the same device (Figure 10). These systems can be used as separate units for one family houses and also in series connection with an insulated storage tank inside the building, regarding larger size applications.

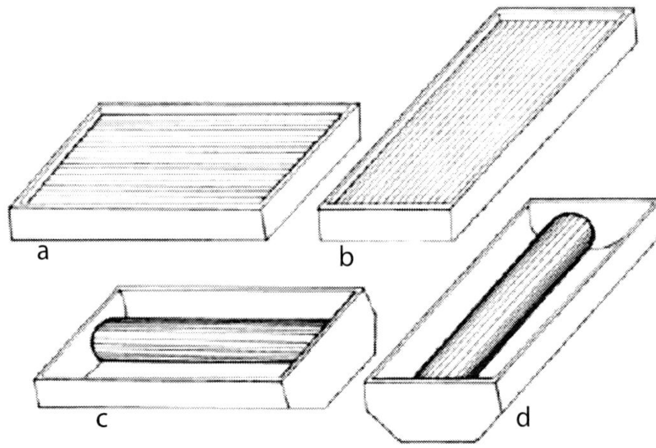

Figure 10. ICS systems with flat and tubular storage tank [27].

II) Solar Collectors with Colored Absorbers

The solar collectors with absorbers of different color than black could be an interesting solution for the wider application of solar energy systems. These colored collectors have lower absorptance and hence operates with lower thermal efficiency than that of the usual black type collectors but they are more interesting to architects for applications on traditional or modern buildings. Regarding cost, an additional amount of about 20% to increase the collector area, can be considered to overcome their lower thermal output compared to the collectors with black absorber of the same type. The application of blue collectors in white buildings in islands, or red brown collectors in buildings with inclined roof and traditional architecture and of other color on modern buildings, could contribute to a wider use of solar thermal collectors [27, 29].

III) Solar Collectors with Booster Reflectors

There are many buildings which have horizontal roof and solar collectors can be installed in parallel rows, placed at a proper distance, in order to avoid collector shading during winter. The space between the parallel rows can be used to provide additional solar radiation on the collector aperture surface by placing booster reflectors from the collector top of the one row to the collector bottom of the next row. These reflectors can contribute to the increase of the thermal energy output by about 20%-50% and from spring to fall this type of installation is suitable for collector operation in higher temperatures, adapting therefore space cooling requirements [27, 30].

IV) Unglazed Solar Collectors

The optical annoyance from reflected light, the use of glass coatings that diffuse reflected light is the most appropriate solution. More often and to avoid the problems from the glare, it is considered worthy to use uncovered (unglazed) solar collectors for low temperature applications (swimming pools, preheating of water, etc). The unglazed solar collectors can be alternative to typical collectors with glazing, for water heating up to about 35°C (Figure 11). In this perspective, uncovered collectors can be combined with colored surface for an even more interesting integration of solar collectors in facades and inclined roofs of buildings [31].

V) Hybrid Photovoltaic/Thermal (PV/T) Systems

The solar energy systems that provide electricity and heat simultaneously are the hybrid photovoltaic/thermal (PV/T) systems, which are consisted of PV modules coupled to water or air heat extraction devices achieving a higher energy conversion efficiency of the absorbed solar radiation [32]. Such installations can contribute to building natural ventilation, operating as solar chimneys on building facade and inclined roof as shown in Figure 12 [32]. A novel PV-Trombe wall is shown in Figure 12 [32].

The PV module provide electricity and also chimney effect which can be utilized to heat the room in winter by recirculation of room air through vents. The same PV-Trombe wall can cool the room by providing chimney effect to circulate the outside air to carry away the heat gain and thereby cooling the room air through induced ventilation in the chimney as shown in Figure 12. The arrangemnent can be made into the PV-Trombe wall by inducing the flow of cool air into the room from shadow side of the room and then passing it through the chimney which also provides cooling effect inside the room.

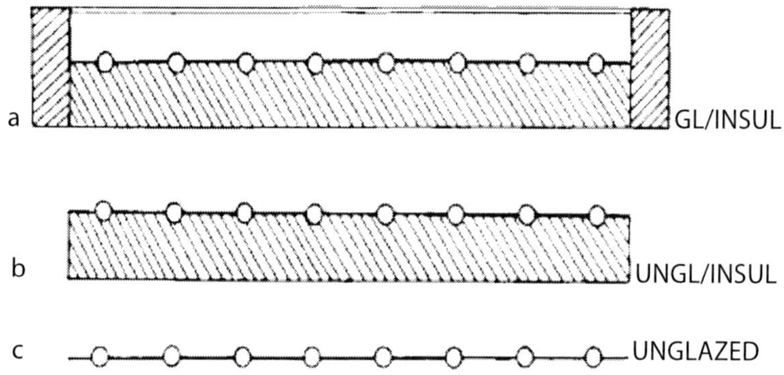

Figure 11. Solar collectors: a) glazed insulated, b) unglazed insulated, c) unglazed [27].

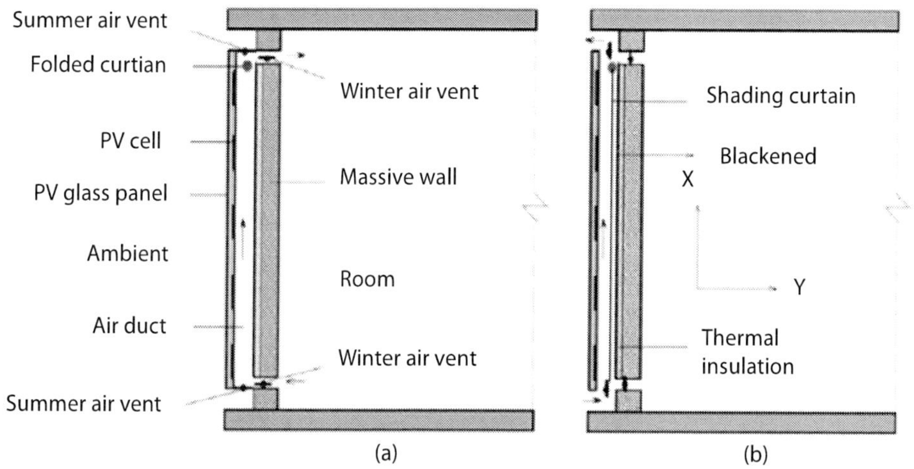

Figure 12. PV-Trombe wall for (a) winter heating and (b) summer cooling [32].

VI) Fresnel Lenses for Building Atria

Fresnel lenses are optical devices for solar radiation concentration and are of lower volume and weight, smaller focal length and lower cost, compared to the thick ordinary lenses. The advantage to separate the direct from the diffuse solar radiation makes Fresnel lenses suitable for illumination control of building interior space, providing light of suitable intensity level and without sharp contrasts. The Fresnel lens concept has been suggested for solar control of the buildings to keep the illumination and the interior temperature at the comfort level [33]. Laboratory scale experimental results give an idea about the application of this new optical system. The collection of 60-80% of the transmitted solar radiation through the Fresnel lenses on linear absorbers leaves the rest amount to be distributed in the interior space for the illumination and thermal building needs. In low intensity solar radiation, the absorber can be out of focus, leaving all light to come in the interior space and to keep the illumination at an acceptable level. The Fresnel lenses can be combined with thermal, photovoltaic, or hybrid type photovoltaic/ thermal absorbers to collect and extract the concentrated solar radiation in the form of heat, electricity or both.

VII) Building Integration of Solar/Wind Systems

Another interesting subject that has been studied is the building integration of solar and wind energy systems.

Both of these devices appear to be the most interesting among renewable energy sources for the built environment. The facades and the horizontal or inclined roofs of houses, hotels, athletic centers and buildings of other types are appropriate surfaces for the application of solar energy conversion systems as they are the photovoltaic panels and the thermal collectors for the electricity and heat demand respectively.

Apart of them small wind turbines (WT) can be mounted on building roofs, mainly at locations with satisfactory wind velocity potential. The use of small WT to typical grid connected residential buildings, hotels, etc, is recently suggested to provide electricity supply and they are suitable mainly for decentralized applications. In addition, they can be effectively combined in complementary operation with photovoltaic and solar thermal collectors [34]. In this direction, there are various systems developed, such as wind turbines of horizontal (HAWT) or vertical (VAWT) axis, wind concentrators, etc. Hybrid photo-voltaic-wind and/or Diesel systems can offer great abilities in the production of energy based on solar and wind energy. In regions where sunshine and wind conditions are good, like the Greek islands, the combined use of photovoltaic and wind turbines has great results for most of the day-night period and also for a very large period of a year, with the thermal collectors covering thermal needs all year round.

The combination of photovoltaic/thermal (PV/T) and WT systems has been suggested as a new concept and the multiple energy conversion systems are the hybrid wind (electric) or solar (electric and thermal) systems (i.e WT or PV/T). They are considered suitable in rural and remote areas with electricity supply from stand-alone units or mini-grid connection. WT or PV/T systems can also be used in typical grid connected applications [35]. For stand-alone WT and PV systems, diesel generators are used in case that solar and wind energies are not sufficient to cover the electrical load.

A very important issue that makes hybrid systems interesting solution for the production of energy is the complementary function of the photovoltaic panels and the wind turbine. PV panels can be useful only in daytime and under a certain solar radiation. On the other hand,

WT can produce energy only when the wind velocity is above a certain rate. So, PV and WT systems can effectively be combined with PV for sunny days and WT for windy nights or for the cloudy days. Also, during winter, where the solar radiation is generally at a low rate, PV systems cannot reach a sufficient performance while WT can offer a lot to energy supply. During summer months, PV systems have fascinating results, which could compensate the unsteady performance of the WT.

Hence, new solar energy systems of better view than the usual types, such as the integrated collector storage (ICS) solar water heaters and the thermal collectors with colored absorbers are suggested for a wider and more aesthetic application of solar energy to buildings. The horizontal roofs of buildings can be effectively used for the conversion of the incoming solar radiation if booster reflectors are mounted between the parallel rows of the collectors. Lighting and temperature control of atria or other interior building spaces can be achieved with Fresnel lenses combined with linear multifunctional absorbers. Alternatively to the building integrated photovoltaic, new solar devices, the hybrid photovoltaic/thermal (PV/T) systems, can be used to provide electricity and heat. The wind energy systems, such as the small wind turbines with horizontal or vertical axis can be considered interesting RES devices for the buildings, which can be effectively combined with PV or PV/T solar systems.

3. Economics of Renewable Energy Technologies

3.1. Economics of PV System for Building

The unit cost of electricity for the typical PV system which can be used for building is summarized in Table 2 [36]. Several values of installed cost in dollars per peak watt were considered along with three different values of minimum attractive rates of return (or discount rate (d)) [36].

Table 2. Unit cost of electricity from photovoltaic technology for building [36]

$/Watt	Unit cost of PV electricity ($/kWh)		
	d = 5%	d = 10%	d = 15%
20	0.92	1.25	1.64
16	0.73	1.00	1.31
12	0.55	0.75	0.98
8	0.37	0.50	0.66
4	0.18	0.25	0.32
2	0.09	0.12	0.16
1	0.04	0.06	0.08

In all these calculations, its was assumed that (i) the useful life of the PV system is 20 years (ii) the annual maintenance cost is 2 percent of the total capital cost of the system, (iii) taxes and insurance costs are not to be paid and (iv) capacity utilization is 25 percent [36].

The optimum size and cost for PV system components (such as PV array, battery bank, inverter andd solar charge controller) with respect to number of full-susnhine hours on the optimally titled surface corresponding to the location of site in the world map was reported by eminent scientists [37].

The optimal tilt for PV array was reported equal to the latitude of the place. The seasonal varioation of this optimum tilt angle for winter and summer months can be considered in the range of latitude ± 15°, +15 considered during winter months and -15 during summer months [37].

3.2. Economics of Wind Power System for Building

Table 3 presents the economic numerical calculation results for wind power generation. The capital cost as well as capacity utilization affect the unit cost electricity from wind turbine power generation systems. At higher values of discount rate, capital cost accounts for 80 % of the total cost of wind generated electricity [36]. Thus, reduction in the capital cost of wind turbines will have a direct impact on the unit cost of electricity generated from the wind resource. In brief, the wind electricity cost can be reduced by (i) reduction in capital cost and (ii) increased energy output as a result of improved aerofoil, larger turbines and the introduction of more efficient operating strategies. The capital cost will be reduced as a result of innovations which may include the use of lighter weight materials and designs leading to lower manufacturing costs.

Table 3. Unit cost of electricity for wind power generation (operation and maintenance cost = 2 % of capital cost, useful life = 20 years, discount rate = 12 %) [36]

Capital cost ($/kW)	Unit cost of electricity ($/kWh) from wind power system for different values of capacity utilization factor (CUF)		
	CUF=0.20	CUF=0.25	CUF=0.30
500	0.044	0.035	0.029
600	0.053	0.042	0.035
700	0.062	0.050	0.041
800	0.070	0.056	0.047
900	0.079	0.063	0.052
1000	0.088	0.070	0.058
1100	0.097	0.078	0.065
1200	0.106	0.085	0.071

3.3. Economics of Solar Hot Water System for Building

There are different conventional options for obtaining the useful thermal energy and their cost comparison is given in Table 4(a) [36]. The efficiency of fuel utilization and unit price of fuel used for water heating have maximum effect on the unit cost of useful thermal energy delivered. The unit cost of useful thermal energy delivered by solar hot water system is presented in the Table 4(b) [36]. As expected, the unit cost of useful thermal energy delivered critically depends upon the total number of days when the solar heated water is delivered to or used by the investor. The unit cost of solar water heater is comparable to that of the conventional cheapest water heating options for high values of capacity utilization and/or low values of discount rate (d).

3.4. Economics of Solar Water Pumping System for Building

The effect of water demand and pumping head on the unit cost of water is given in Table 5 [36]. The solar PV surface pump was used for maximum of 10000 m^3 water demand at 10 m head while for all water demands of 20 m head were met by using the submersible solar PV water pump.

Table 4(a). Unit cost of useful thermal energy for water heating systems [36]

Fuel	Unit	Calorific Value (MJ/unit)	Efficiency of Utilization	Unit Price of Fuel (Rs./unit)	Unit cost of useful thermal energy (Rs./MJ)
Electricity	kWh	3.6	0.9	3	0.93
	kWh	3.6	0.9	5	1.54
LPG	kg	45	0.6	16	0.59
	kg	45	0.6	25	0.93
Kerosene	kg	45	0.4	7.5	0.42
	kg	45	0.4	14	0.78
Fuelwood	kg	16	0.1	1.5	0.94
	kg	16	0.25	1.5	0.38
Natural Gas	kg	39	0.6	2	0.08
	kg	39	0.6	2.5	0.11

Table 4(b). Unit cost of useful thermal energy for solar water heating system [36]

Capital cost (Rs.)	Capacity Utilization	Unit cost of useful thermal energy delivered (Rs./MJ)			
		d = 6%	d = 8 %	d = 10 %	d = 12 %
15000	0.1	2.55	2.87	3.21	3.56
	0.2	1.27	1.43	1.60	1.78
	0.3	0.85	0.96	1.07	1.19
	0.4	0.64	0.72	0.80	0.89
	0.5	0.51	0.57	0.64	0.71
	0.6	0.42	0.48	0.53	0.59
	0.7	0.36	0.41	0.46	0.51
	0.8	0.32	0.36	0.40	0.45
	0.9	0.28	0.32	0.36	0.40
20000	0.1	3.40	3.82	4.28	4.75
	0.2	1.70	1.91	2.14	2.38
	0.3	1.13	1.27	1.43	1.58
	0.4	0.85	0.96	1.07	1.19
	0.5	0.68	0.76	0.86	0.95
	0.6	0.57	0.64	0.71	0.79
	0.7	0.49	0.55	0.61	0.68
	0.8	0.42	0.48	0.53	0.59
	0.9	0.38	0.42	0.48	0.53

Table 5. Unit cost of water pumping for different head [36]

Annual demand of water (m^3)	Type of pump set	Unit cost of water (Rs/m^3) for the head of		
		5 m	10 m	20 m
7000	Solar PV	0.907	2.640	5.17
	Windmill	0.612	0.830	Not applicable
	Producer Gas Diesel	0.885	1.131	1.375
	Biogas Diesel	0.562	0.728	0.847
	Diesel Engine	0.610	0.848	1.113
	Electric Motor	0.493	0.563	0.650
10000	Solar PV	0.882	2.610	Not applicable
	Windmill	0.428	Not applicable	Not applicable
	Producer Gas Diesel	0.656	0.864	1.103
	Biogas Diesel	0.411	0.545	0.664
	Diesel Engine	0.467	0.637	0.938
	Electric Motor	0.359	0.420	0.507

As expected, increase in water demand results in reduced cost of pumping water for all resource technology combinations reported in Table 5 [36]. The reason behind this is improvement in the capacity utilization for all resource technology combinations except solar PV pumps.

3.5. Economics of Solar PV Lanterns for Building

In the commercial street markets in developing countries, the use of solar lanterns has increased tremendously because of low cost option without any pollution as compared with conventional (petromax or kerosene) lamps in the street markets. The cost comparison was made based on the January 1996 price (1 US $ = Rs. 35). These solar PV lanterns need to be promoted for rural and semi-urban areas in developing countries like India, China, South Africa etc. Two hurricane lanterns are needed to provide illuminance equivalent to 3 W solar PV lanterns for identical conditions [36]. While the light output of 11 W solar PV lantern is comparable to a petromax. In view of this, the cost per hour of illumination of 3W solar PV lantern was compared with cost per hour of illuminance from two hurricane lanterns. Similarly, cost comparison was made between 11 W solar PV lantern and petromax. The cost comparison results are given in Table 6. The market price of kerosene was considered as Rs. 8/litre (January 1996) [36]. Hence, in the present case, these solar lanterns are more economical because of rise in fossil fuel prices.

Table 6. Cost per hour of illumination of solar PV lanterns [36]

	Two Hurricane Lanterns	3 W Solar PV Lantern	Petromax Lantern	11 W Solar PV Lantern
Cost of Lighting (Rs./hr)	0.86	0.23	0.73	0.61

Table 7. Cost comparison of distillation systems for fresh water production [36]

Sr.No.	Distillation Techniques nnnn	Distilled water cost, d=10% (US $/m^3)
A.	Conventional Distillation Methods	
1.	Single Purpose, Multistage Flash Distillation	1.74
2.	Thermal Vapour Compression	1.02
3.	Single Purpose, Long Tube, Multiple Effect Distillation	1.05
4.	Dual Purpose, Long Tube, Multiple Effect Distillation	0.61
5.	Seawater Reverse Osmosis	0.71
B.	Solar Assisted Distillation Methods	
6.	Solar Pond, Long Tube, Multiple Effect Distillation	0.80-0.88
7.	Solar High Temperature Parabolic Mirrors-Long Tube Multiple Effect Distillation	0.87-0.91
8.	Parabolic mirrors, Thermal Vapour Compression	1.32-1.44
9.	Solar High Temperature Parabolic Mirrors, Seawater Reverse Osmosis	0.88-0.90

3.6. Economics of Solar Distillation for Drinking Water Supply in Arid Area Buildings

The conventional solar stills are the simplest devices for providing distilled water in the remote arid areas with little rainfall and fresh water sources. The conventional distillation using fossil fuels are more complicated than solar distillation systems.

The conventional systems require capital amortization, maintenance and operation cost. The unit cost of distilled water produced by conventional distillation systems will depend upon the above factors besides the cost of fuel used. The solar assisted distillation systems are similar to the conventional desalination systems in all respects except for the energy source and associated solar energy collection sub-systems. Solar assisted distillation systems are even more complicated in operation than conventional plants due to its dependence on a variable input source.

In principle, other renewable energy sources such as wind energy, biomass energy and ocean thermal energy etc. can be used for distillation. A comparison of the estimated distilled water cost produced from various conventional distilled methods with solar assisted distillation techniques is presented in Table 7 [36].

4. Economics of Solar Lighting and Passive Heating/Cooling Techniques for Building

There are various passive solar heating/cooling and solar lighting techniques. The economics of these systems are briefly discussed below:

4.1. Economics of Daylighting in Buildings

There are various daylighting techniques used for natural lighting of interior space of buildings to cut down the artificial lighting energy requirement of the building during daytime.

There is court yard in the Indian Habitat Centre in New Delhi (India) to serve the purpose of employ relaxation with fountain and natural beauty of in-house plants. It is shaded with transparent colored plastic sheets so that it allows light intensity which is desired for normal illuminance for relaxation. Similarly, in TERI retreat building in New Delhi (India), semi transparent PV modules are placed over the stair case roof to provide natural light required for walking on the stair case.

This daylighting system leads to artificial lighting energy savings. Consider a skylight building located at Solar Energy Park, IIT Delhi (India) which provides an average ammount of 400 Lux (or Lumens/m^2) on the floor throughout the year. This is equivalent to 10000 lumens for 25 m^2 of floor area. It is know that the average efficacy of energy efficient 15 W CFL lamp is 40 Lumen/W. Hence, the natural skylight is equivalent to 17 CFL lamps of 15 W. If natural light is available for 6 h/day then daily energy saving from 17 CFL lamps is calculated as 1.53 kWh/day. The annual energy saving from this natural skylight in the building is calculated as 559 kWh/year. If the cost of electricity is Rs.5/kWh, then the annual ammount of reduction in electricity bill can be determined as Rs. 2795/year.

If the cost incurred in retrofitting of skylight is Rs.14000 then the simple payback period can be estimated nearly 5 years. Hence, it is economical to retrofit skylight in the existing building for obtaining natural light inside the room. However, the availability of skylight inside the room varies with time and day of the year for the particular location. Hence, it is desired to analyze the availability of natural light and orientation of building for the particular location in the world before utilizing skylight in building. There are lot of research in this area to design skylight based on the location of building in the world i.e. climate specific building design or also know as solar buildings.

4.2. Economics of Retrofitting Trombe Wall for Passive Heating in Buildings

Based on the case study of Trombe wall for passive heating of building used for honey storage in winter months [38], the following results were obtained.

Before Retrofit of Building

Equipments: Air heater (oil filled radiator) 2.3 kW and Desert cooler 746 W (1 HP)

- Annual energy consumption of the building is 4404 kWh/year
- Comprises of 3312 kWh/year for air heater and 1092 kWh/year for desert cooler
- Energy intensive honey storage building
- The annual electricity bill to the stakeholder is Rs.13212/year for cost of electricity Rs.3/kWh

After Retrofitting of the Building with Trombe Wall

Equipments: Trombe wall and Desert cooler:

- Annual energy consumption reduce to 1092 kWh/year (only for desert cooler)
- Large saving of energy 3312 kWh/year for heating room air in winter
- Electricity bill Rs.3276/year
- Saving of electricity bill of Rs.9936/year after retrofit using Trombe wall
- Total Trombe wall retrofit cost Rs.10000 (approximately) in India
- Simple payback period is less than 1 year

Based on this case study, the passive heating concepts are most suitable for cold climatic conditions in the world e.g. Canada, Europe etc. and in India e.g. Leh, Laddakh regions etc.

4.3. Economics of Integrating Earth to Air Heat Exchanger (EAHE) for Passive Heating/Cooling of Buildings

There are lots of case studies of integrating Earth to air heat exchangers with buildings for space heating/cooling in buildings all over the world. One of the case studies discussed here is based on the EAHE integrated with adobe house for heating/cooling of building in New Delhi composite climatic condition in India [39]. The results of this study showed that annual energy saving potential of adobe house before and after integration of EAHE with the renovated adobe house are 4946 kWh/year and 10321 kWh/year respectively [39]. This shows that there is considerable energy saving potential of adobe house integrated with EAHE as compared to normal adobe house thermal performance. The life cycle cost analysis showed that the integration of EAHE with adobe house has substantial benefits with payback period less than 2 years for 16 % interest rate on the capital investment made on EAHE. The additional annual energy saving potential of EAHE was estimated as 5376 kWh/year comprising of heating and cooling energy saving potential of 3012 kWh/year and 2364 kWh/year respectively corresponding to New Delhi climatic data [39]. The monetary annual cash inflow of EAHE was calculated as €448/year while annual cash outflow for operation and maintenance was €240.9/year. The annual mitigation of CO_2 emissions into the environment with and without integration of EAHE were estimated nearly 16 tons/year and 8 tons/year and corresponding carbon credits of €320/year and €160/year respectively [39].

Conclusion on Building and Sustainable Development

Modern day buildings are highly energy intensive with a significant consumption of energy right from the construction phase to the operation and maintenance stage, owing to global energy crisis suitable strategies need to be developed to address energy conservation in buildings.

There are various methods to reduce the use of conventional energy from fossil fuels to meet the energy requirement for the building. The combination of various solar passive design aspects can easily be integrated in new buildings based on the site, orientation of building and local climatic conditions. The retrofit of Trombe wall as solar passive heating design concepts in honey storage building had shown promising results for winter heating. Similarly, the use of proper design of daylighting can leads to tremendous reduction in use of artifical lights during daytime and thereby reduces the energy consumption by building for lighting. Hence, integration of solar passive features into the building leads to reduction in energy consumption of building which ultimately reduces the CO_2 emissions and helps in sustainable development. The second important aspect is the use of low embodied energy and locally available building materials for building construction to avoid tremendous energy input in construction of building and thereby reduces the CO_2 emission from the building sector. The third aspect is to reduce energy consumption of building for operation using energy derived from fossil fuels. So, as a feasible alternative the focus is on the promotion of renewable energy technology in meeting building energy requirements. When the building energy is completely met by renewable energy system then it is known as a highly energy efficient or zero emission green building. The economics of various renewable energy systems is given for the acceptance of these technologies as compared to the conventional energy sources.

All the European Union member countries have a very critical target to meet from the renewable source of energy. The target is to meet 12% of total energy consumption from renewables out of which 21% of electricity to be met from renewable energy systems up to 2010 [27]. Hence, there is urgent need to develop policies all over the world for building integrated renewable energy systems.

References

[1] N. Zografakis, Technologies for rational use and savings of energy in buildings, *Energy* 62 (November–December 2000): 112–114 (in Greek).

[2] D.L. Loveday, K.C. Parsons, A.H. Taki, S.G. Hodder, Displacement ventilation environments with chilled ceilings: thermal comfort design within the context of the BS EN ISO7730 versus adaptive debate, *Energy and Buildings* 2002; 34(6): 573–579.

[3] S. Heidari, S. Sharples, A comparative analysis of short-term and long-term thermal comfort surveys in Iran, *Energy and Buildings* 2002; 34(6): 607–614.

[4] S. Yannas, Solar Energy and Housing Design, Volume-1, *Principles, Objectives, Guidelines,* Architectural Association, London, 1994.

[5] T. Markis, J.A. Paravantis, Energy conservation in small enterprises. *Energy and Buildings,* 2007; 39(4): 404–415.

[6] Dorota Chwieduk, Towards sustainable-energy buildings. *Applied Energy*, 2003; 76(1-3): 211–217.

[7] *http://www.santabarbaraca.gov/NR/rdonlyres/BFD34004-668E-4238-B7AB-DE8C2F05C648/0/Exhibit_B_Passive_Solar_Guidelines_ Recognition_Program.pdf,* accessed on February 5, 2009.

[8] *www.nrel.gov/docs/fy01osti/29236.pdf*, accessed on November 15, 2008.

[9] *http://www.santabarbaraca.gov/Resident/Home/Forms/Building_and_Safety.htm,* accessed on February 9, 2009.

[10] *http://www.greenbuilder.com/sourcebook/PassSolGuide1-2.html#heat,* accessed on November 15, 2008.

[11] *http://www.azsolarcenter.com/technology/pas-2.html,* accessed on November 16, 2008.

[12] *http://en.wikipedia.org/wiki/Passive_solar_building_design#cite_note-16*, accessed on November 17, 2008.

[13] *http://www.eere.energy.gov/consumer/your_home/space_heating_cooling/index.cfm/my topic=12350,* accessed on November 18 2008.

[14] *http://en.wikipedia.org/wiki/Solar_chimney,* accessed on November 212008.

[15] *http://www.greenbuilder.com/sourcebook/PassSolGuide3.html,* accessed on November 15, 2008.

[16] *http://en.wikipedia.org/wiki/Solar_air_conditioning*, accessed on November 29, 2008.

[17] Catarina Thormark. A low energy building in a life cycle—its embodied energy, energy need for operation and recycling potential. *Building and Environment*, 2002; 37(4): 429 -435.

[18] Alcorn, J. (1995). *Embodied energy coefficients of building materials*. Centre for Building performance research, Victoria University of Wellington.

[19] B.V.Venkatarama Reddy and K. S. Jagadish. Embodied energy of common and alternative building materials and technologies, *Energy and Buildings*, 2003; 35(2): 129-137.

[20] *http://www.architecture2030.org/regional_solutions/materials.html.*

[21] *Building materials in India: 50 years*. T N Gupta. Building Materials and Technology Promotion Council, Ministry of Urban Affairs and Employment, Government of India, New Delhi. 1998.

[22] Arvind Chel and G.N. Tiwari. Thermal performance and embodied energy analysis of a passive house – case study of vault roof mud-house in India. *Applied Energy*, 2009; 86(10): 1956-1969.

[23] *http://www.iea.org/textbase/nppdf/free/2000/Renew_main2003.pdf*, accessed on December 10, 2008.

[24] *http://en.wikipedia.org/wiki/Renewable_energy#cite_note-REN21-2007-0*, accessed on December 11, 2008.

[25] *http://www.ren21.net/pdf/RE2007_Global_Status_Report.pdf*, accessed on December 12, 2008.

[26] *http://www.undp.org/energy/activities/wea/drafts-frame.html,* accessed on December 10, 2008.

[27] *http://www.inive.org/members_area/medias/pdf/Inive%5CMilos2006%5C46_Tripanag nostopoulos_6P.pdf*, accessed on December 16, 2008.

[28] Tripanagnostopoulos Y., Souliotis M., ICS solar systems with horizontal E-W and vertical N-S cylindrical water storage tank. *Renewable Energy* 2004; 29(1): 73-96.

[29] Kalogirou S., Tripanagnostopoulos Y. and Souliotis M., Performance of solar systems employing collectors with colored absorber, *Energy and Buildings* 2005; 37(8), 824–835.

[30] Tripanagnostopoulos Y. and Souliotis M., 2005. Booster Reflector Contribution to Performance Improvement of Solar Collectors. *Proceedings in CD, International Conference WREC 2005*, Aberdeen, UK, 22–27 May: 63–68.

[31] Tripanagnostopoulos Y., Nousia T. and Souliotis M., 2000. Low cost improvements to building integrated air cooled hybrid PV-Thermal systems *Proceedings of 16th European PV Solar Energy Conference*, Glasgow, U.K. 1-5 May: 1874-1877.

[32] Ji J., Yi H., He W., Pei G., PV-Trombe wall design for buildings in composite climates, *Journal of Solar Energy Engineering by ASME* 2007; 129: 431-437.

[33] Tripanagnostopoulos Y., Siabekou Ch. and Tonui J.K., The fresnel lens concept for solar control of buildings. In *Proceedings of International Conference. PALENC-2005,* Santorini, Greece 2005: 977-982.

[34] Tripanagnostopoulos Y., Tselepis S., Hybrid solar/ wind PVT/WT building integrated systems. In *Proceedings of 2nd International European PV-Hybrid and Mini-Grid Conference;* Kassel, Germany, 25-26 September, 2003: 329-333.

[35] Tripanagnostopoulos Y., Tselepis S., Souliotis M., Tonui J. K. and Christodoulou A., 2004. Practical aspects for small wind turbine applications. *Proceedings of International EWEC 2004 Conference,* Track 7; Autonomous and/or Hybrid Systems, London, UK, 22–25 November.

[36] Kandpal T.C. and Garg H.P., *Financial Evaluation of Renewable Energy Technologies*, Macmillan India Ltd., (2003): 149-363.

[37] Arvind Chel, G.N.Tiwari and Avinash Chandra, Sizing and cost estimation methodology for stand-alone residential PV power system, *International Journal of Agile Systems and Management*, 2009; 4(1/2): 21-40.

[38] Arvind Chel, J.K. Nayak and Geetanjali Kaushik, Energy conservation in honey storage building using Trombe wall, *Energy and Buildings*, 2008; 40(9): 1643-1650.

[39] Arvind Chel and G.N. Tiwari, Performance evaluation and life cycle cost analysis of earth to air heat exchanger integrated with adobe building for New Delhi composite climate, *Energy and Buildings*, 2009; 41(1): 56-66.

In: Carbon Dioxide Emissions: New Research
Editors: M. Carpenter and E. J. Shelton

ISBN: 978-1-62257-436-0

Chapter 7

ALLOCATING CARBON DIOXIDE EMISSIONS FROM COGENERATION: AN EXERGY-BASED APPROACH

Marc A. Rosen*
Faculty of Engineering and Applied Science
University of Ontario Institute of Technology,
Ontario, Canada

ABSTRACT

The allocation of carbon dioxide emissions from cogeneration is often challenging, as existing methods are not soundly based, inconsistent, overly complex and difficult to apply. The author proposes that exergy methods can form the basis for rationally and meaningfully allocating cogeneration-based carbon dioxide emissions, and describes and illustrates such methods in this chapter and compares them with other methods. The results suggest that the exergy-based approach provides a sensible basis for a meaningful overall approach for carbon dioxide emissions trading. It is concluded that the exergy-based method for allocating carbon dioxide emissions from cogeneration is rational, useful and superior to other methods. The results aid designers of energy systems, and decision and policy makers in companies and government, especially regarding how emissions should be reduced, how and where cogeneration should be used, and how such carbon dioxide emissions can be traded.

1. INTRODUCTION

The potential benefits of cogeneration (i.e., combined heat and power) are significant. But many companies, government agencies and researchers have struggled with the question of how to allocate carbon dioxide emissions for cogneration since it has multiple products. Some work has been done, e.g., there exist efficiency, work-potential and heat-content methods (Strickland and Nyboer, 2002a, 2002b; Upton, 2001; Phylipsen et al., 1998), but the

* Corresponding author: Email: marc.rosen@uoit.ca, Tel: 905/721-8668

results are not soundly reasoned, not universally accepted, often conflicting, overly complex and difficult to apply. The author feels that exergy methods can form the basis of sound and meaningful allocation methods for CO_2 emissions. Such exergy-based methods are described and compared with the other allocation methods in this chapter. Illustrative examples are presented, and a related approach for CO_2 emissions trading is described.

Much work on assessing the performance of cogeneration systems and their emissions has been carried out in Europe. For instance, the European Parliament (2004) issued a directive in February 2004 on the promotion of cogeneration based on a useful heat demand in the internal energy market. Also, the European Committee for Standardization (CEN) and the European Committee for Electrotechnical Standardization (CENELEC) published in 2004 a workshop agreement manual for the determination of combined heat and power (CEN/CENELEC, 2004). Although these two documents have been criticized (Verbruggen, 2007a, 2007b), they are widely accepted for primary energy savings for cogeneration. They provide indicators for carbon dioxide emissions allocation from cogeneration systems. However, these approaches ignore exergy, and thus suffer from the same difficulties as mentioned earlier of allocation methods.

2. COGENERATION

Cogeneration usually refers to the simultaneous production of two energy forms (electricity, and heat in the form of steam and/or hot water) from one energy source (often a fossil fuel). A cogenerator can be a utility, an industry, a government, or any other party. Cogeneration has been used, particularly by industry, for over a century.

Cogeneration systems are often extensions of thermal electricity-generation systems. In thermal electrical generating stations, the energy content of a resource (normally a fossil fuel) is converted to heat (in the form of steam or hot gases) which is then converted to mechanical energy (in the form of a rotating shaft), which in turn is converted to electricity. A portion (normally 20 to 45%) of the heat is converted to electricity, and the remainder is rejected to the environment as waste.

Cogeneration systems are similar to thermal electricity-generation systems, except that a percentage of the generated heat is delivered as a product, normally as steam or hot water, and the quantities of electricity and waste heat produced are reduced. Overall cogeneration efficiencies based on both the electrical and thermal energy products of over 80% are achievable. Other advantages generally reported from cogenerating thermal and electrical energy rather than generating the same products in separate processes include:

- reduced energy consumption,
- reduced environmental emissions (due to reduced energy consumption and the use of modern technologies in large, central installations), and
- more economic, safe and reliable operation.

Most thermal systems for large-scale electricity generation are based on steam and/or gas turbine cycles, and can be modified relatively straightforwardly for cogeneration.

Two main categories of heat demands can usally be satisfied through cogeneration:

(i) residential, commercial and institutional processes, which require large quantities of heat at relatively low temperatures (e.g., for air and water heating); and
(ii) industrial processes, which require heat at a wide range of temperatures (e.g., for drying, heating and boiling in, for instance, chemical processing, manufacturing, metal processing, mining and agriculture).

The use of a central heat supply to meet residential, commercial and institutional heat demands is often referred to as district heating. As well as satisfying heat demands, cogenerated heat can drive chillers, a particularly beneficial application where the annual peak electrical demand is associated with the summer cooling load.

Many general descriptions and studies of cogeneration systems have been reported (MacRae, 1992; Rogner, 1993; FVB/Eltec, 1993; MacLaren, 1988; Henneforth and Todd, 1988; Acres, 1987; Horlock, 1987; Rosen, 1993, 1994, 1998; Rosen et al., 1997; Hart and Rosen, 1994; Rosen and Le, 1994). Cogeneration systems are in use throughout the world (e.g., thousands are listed by the Association of Energy Engineers), and the basic technology is proven. Numerous examples exist of large cogeneration systems.

The size and type of a cogeneration system are normally selected to match as optimally as possible the thermal and electrical demands. Many matching schemes can be used. Systems can be designed to satisfy the electrical or thermal base-loads, or to follow the electrical or thermal loads. Storage systems for electricity (e.g., batteries) or heat (e.g., hot water or steam tanks) are often used to overcome periods when demands and supplies for either electricity or heat are not coincident. Cogeneration systems are sometimes used to supply only the peak portions of the electrical or thermal demands.

3. Exergy Methods

3.1. Exergy

The exergy of an energy form or a substance is a measure of its usefulness or quality. Exergy is based on the first and second laws of thermodynamics, and combines the principles of conservation of energy and non-conservation of entropy. Exergy is defined as the maximum amount of work which can be produced by a system or a flow of matter or energy as it comes to equilibrium with a reference environment. Exergy is a measure of the potential of the system or flow to cause change, as a consequence of not being completely in stable equilibrium relative to the reference environment. Unlike energy, exergy is not subject to a conservation law (except for ideal, or reversible, processes). Rather exergy is consumed or destroyed, due to irreversibilities in any real process. The exergy consumption during a process is proportional to the entropy created due to irreversibilities associated with the process. Exergy and exergy analysis are discussed further elsewhere (Dincer and Rosen, 2007; Sato, 2005; Szargut, 2005; Hevert and Hevert, 1980; Kotas, 1995; Moran, 1989; Moran and Sciubba, 1994, 2004; Moran and Shapiro, 2007; Szargut et al., 1988; Szargut, 1980; Edgerton, 1992; Rosen, 1999).

As a simple illustration, consider an adiabatic system containing fuel and air at ambient conditions. The fuel and air react to form a mixture of hot combustion gases. During the combustion process, the energy in the system remains fixed because it is adiabatic. But the exergy content declines as combustion proceeds due to the irreversibilities associated with the conversion of the high-quality energy of fuel to the lower quality energy of combustion gases. The different behaviours of energy and exergy during this process are illustrated qualitatively in Figure 1.

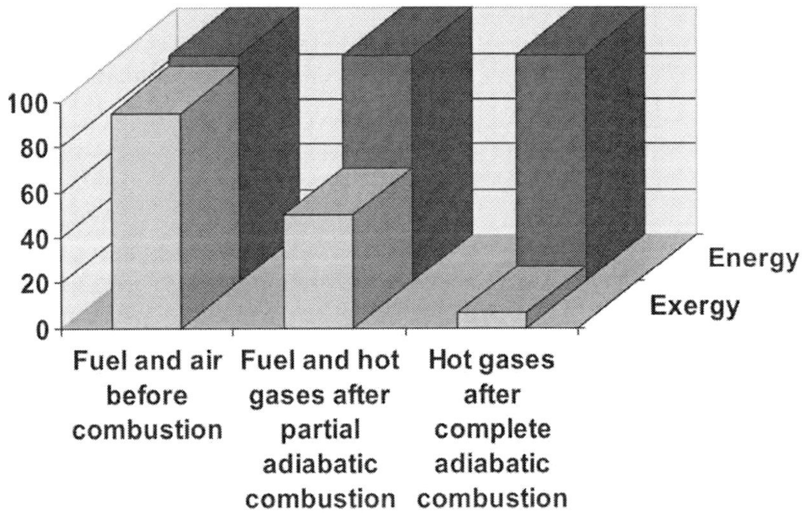

Figure 1. Qualitative comparison of energy and exergy changes during fuel combustion.

3.2. Exergy Analysis

Exergy analysis is a methodology that uses the conservation of mass and conservation of energy principles together with the second law of thermodynamics for the analysis, design and improvement of energy and other systems. The exergy method is useful for improving the efficiency of energy-resource use, for it quantifies the locations, types and magnitudes of wastes and losses. In general, more meaningful efficiencies are evaluated with exergy analysis rather than energy analysis, since exergy efficiencies are always a measure of the approach to the ideal. Therefore, exergy analysis identifies accurately the margin available to design more efficient energy systems by reducing inefficiencies. Many engineers and scientists suggest that thermodynamic performance is best evaluated using exergy analysis because it provides more insights and is more useful in efficiency-improvement efforts than energy analysis. Many applications of exergy analysis to processes and systems have been reported (Dincer and Rosen, 2007; Cownden et al., 2001; Rosen, 1992, 1996, 2000; Rosen and Dincer, 2002; Rosen and Horazak, 1995). Some of those have included investigations by the author of cogeneration using exergy methods (Dincer and Rosen, 2007; Rosen, 1990; Rosen and Le, 1996).

For exergy analysis, the characteristics of the reference environment (temperature, pressure and chemical composition) must be specified. The results of exergy analyses, consequently, are relative to the reference environment, which often is modelled after the actual local environment. The exergy of a system is zero when it is in equilibrium with the reference environment. The tie between exergy and the environment has implications regarding environmental impact has been investigated previously (Dincer and Rosen, 2007; Sciubba, 1999, 2004; Ayres et al., 1998; Connelly and Koshland, 1997; Creyts and Carey, 1997; Berthiaume et al., 2001; Crane et al., 1992; Rosen and Dincer, 1997, 1999, 2001; Gunnewiek and Rosen, 1998; Daniel and Rosen, 2002).

3.3. Exergy of Cogeneration Commodities

It is useful to understand the exergy values associated with electrical and thermal energy. While the energy and exergy contents of electricity are equivalent, these quantities generally differ for thermal energy, sometimes significantly. The variation with temperature of the ratio of exergy to energy associated with thermal energy is presented in Table 1, considering thermal energy transferred at temperatures above the environment temperature (i.e., heat) and below (i.e., cold). For heat, the ratio of exergy to energy varies from zero when the thermal energy is transferred at the environment temperature to unity as the temperature of heat transfer approaches infinity.

Table 1. Comparison of Quality of Various Types of Thermal Energy, for a Reference-environment Temperature of 20°C

Thermal energy type	Temperature (°C)	Exergy-to-energy ratio
Heat	2000	0.871
	1000	0.770
	500	0.621
	200	0.381
	100	0.215
	50	0.093
	20	0.000
Cold	10	−0.035
	0	−0.073
	−10	−0.114
	−50	−0.314
	−100	−0.694
	−200	−3.014
	−250	−11.74

For cold, the values of exergy rate are negative, implying that although heat is taken out of a system to make it colder, the exergy associated with the thermal energy is input to the system to make it colder. That is the flows of energy and exergy in such instances are in opposite directions. This observation implies what is intuitively understood when dealing with systems at below-environment temperatures: cold is the useful commodity. Also, the magnitude of the ratio of exergy to energy for cold varies from zero when the thermal energy is transferred at the environment temperature to greater than unity as the temperature of heat transfer approaches absolute zero. For very cold systems, therefore, the exergy transfer can be larger than the energy transfer.

Thermal energy is often transferred via a medium (often water) in cogeneration. To provide a qualitative comparison of the energy quality, the quality of energy based on exergy of a range of energy forms follows, in descending order:

- Ultra-low-temperature cryogenic cool (≈3 K)
- Electricity, work
- Hydrocarbon fuels
- Very high-temperature heat (≈2000 K)
- High-pressure steam, cryogenic cool (≈200 K)
- Hot water, normal refrigeration cool
- Space heating, space cooling

3.4. Illustrative Example

The use of exergy methods to analyze a device so as to permit its performance to be better understood and its efficiency improved is demonstrated for an illustrative example: electricity generation using a coal-fired steam power plant. The plant considered is the coal-fired Nanticoke generating station, which has been operating since 1981 in Ontario, Canada. Each of the eight units in the station has a net electrical output of 505 MW. A single unit of the electrical generating station is illustrated in Figure 2, and consists of four main sections (Rosen, 2000):

a) *Steam Generators*: Pulverized-coal-fired natural-circulation steam generators combust coal to produce primary and reheat steam. Air is supplied to the furnace by motor-driven forced-draft fans, and regenerative air pre-heaters are used. The flue gas passes through an electrostatic precipitator and exits the plant via multi-flued chimneys.
b) *Turbine Generators and Transformers*: The steam produced passes through a turbine generator, which is connected to a transformer. Each turbine generator has one single-flow high-pressure cylinder, one double-flow intermediate-pressure cylinder and two double-flow low-pressure cylinders. Steam exhausted from the high-pressure cylinder is reheated in the steam generator. Several steam extractions from the turbines preheat feed water in low- and high-pressure heat exchangers and one spray-type open de-aerating heat exchanger. The low-pressure turbines exhaust to the condenser.
c) *Condensers*: Cooling water condenses the steam exhausted from the turbines.
d) *Preheating Heat Exchangers and Pumps*: The temperature and pressure of the condensed steam are increased in a series of pumps and heat exchangers.

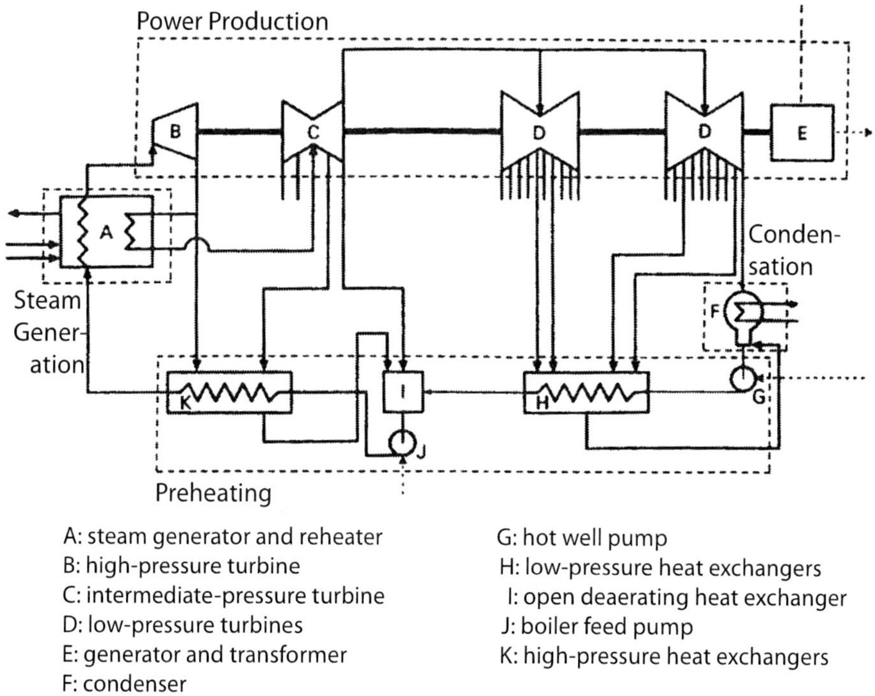

Figure 2. Electrical generating station. The external inputs are coal and air, and the output is stack gas and solid waste for unit A. The external outputs for unit E are electricity and waste heat. Electricity is input to units G and J, and cooling water enters and exits unit F.

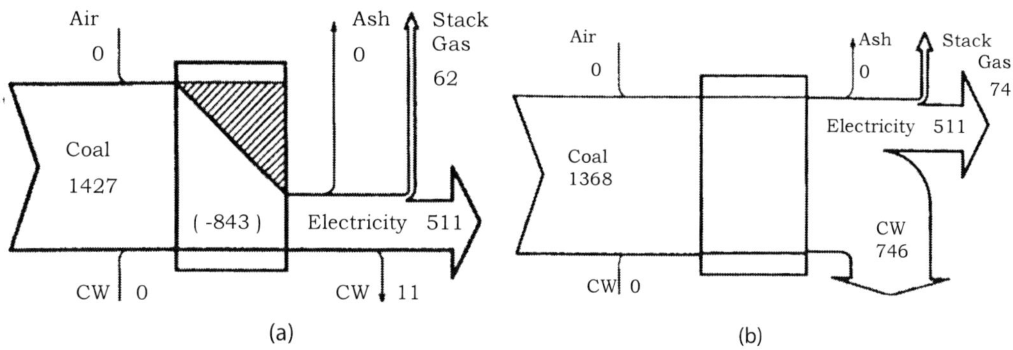

Figure 3. Overall energy and exergy balances for the generating station, represented by a rectangle. Widths of flow lines are proportional to the relative magnitudes of the represented quantities. CW denotes cooling water. a) Exergy balance showing flow rates (positive values) and consumption rate (negative value, denoted by hatched region) of exergy (in MW). b) Energy balance showing flow rates of energy (in MW).

Exergy and energy analyses of the station have been performed (Rosen, 2000). Overall balances of exergy and energy for the station are illustrated in Figure 3, and the main findings, which improve understanding of the thermodynamic behaviour of the plant and help identify areas having significant efficiency-improvement potential, follow:

- The overall-plant energy efficiency (ratio of net electrical energy output to coal energy input), is 37%, and the corresponding exergy efficiency 36%.

- In the steam generators, the energy and exergy efficiencies consider the increase in energy or exergy of the water as the product. The steam generators appear significantly more efficient on an energy basis (95%) than on an exergy basis (50%). Physically, this discrepancy implies that, although most of the input energy is transferred to the preheated water, the energy is degraded as it is transferred. Most of the exergy losses in the steam generators are associated with internal consumptions (mainly due to combustion and heat transfer).
- A large quantity of energy enters the condensers (about 775 MW per unit), of which close to 100% is rejected. A small quantity of exergy enters (about 54 MW for each unit), of which about 25% is rejected and 75% internally consumed.
- In other plant devices, energy losses are very small (about 10 MW total), and exergy losses were found to be moderately small (about 150 MW total). The exergy losses are almost completely associated with internal consumptions.

4. Methods for Allocating CO_2 Emissions for Cogeneration

Some methods have been developed for allocating carbon dioxide emissions from cogeneration systems to the electrical and thermal energy products. The need for these methods is premised on the fact that when the owner of the cogeneration plant, the thermal energy user and the electrical energy user are not the same, a method for allocating the emissions is needed to ensure each party is credited with their appropriate share of the emissions from the system. In addition, having a meaningful allocation method allows the sources of carbon dioxide and other emissions to be better understood and, where appropriate, reduced.

Strickland and Nyboer (2002a, 2002b) list seven methods of calculating the fuel allocation to the thermal and electrical energy products of a cogeneration system. Allocations in those methods are based on product energy contents, product exergy contents, product economic values, incremental fuel consumption to electrical production, incremental fuel consumption to the heat production, shared emission savings between heat and electricity, and agreement. In their work, Strickland and Nyboer (2002a, 2002b) adapt the calculational methods introduced earlier by Phylipsen et al. (1998). In the methods used, the fuel allocation is multiplied by the appropriate carbon dioxide emission factor to evaluate the share of emissions allocated to each product.

Others have also investigated methods for allocating greenhouse gas emissions associated with manufacturing and other industries. Such investigations have been carried out by the World Resources Institute, Washington, DC and the National Council for Air and Stream Improvement, Inc., Corvallis, OR, as evidenced by correspondences between these organizations (Upton, 2001). In general, the allocation methods discussed by Upton (2001) are variations on those discussed by Strickland and Nyboer (2002a, 2002b) and Phylipsen et al. (1998).

The allocation methods presented by Strickland and Nyboer (2002a, 2002b), based on Phylipsen et al. (1998), are reviewed here, after modifying them so that they present the *fractions*, rather than the total, carbon dioxide emissions allocated to each product.

4.1. Allocation Based on Energy Content of Products

Allocations are evaluated in proportion to the energy contents of the products, as follows:

$$f_E = E/(E + Q)$$

$$f_H = Q/(E + Q)$$

where f_E and f_Q denote respectively the fractions of the emissions allocated to the electrical and thermal products, and E and Q denote respectively the net outputs of electrical energy and thermal energy from the cogeneration system. The term Q can represent an actual transfer of thermal energy, or the net thermal energy transferred via a material flow in and out of a heat exchange device. Although this allocation method is straightforward and simple, it ignores the quality of energy and focuses only on the quantities involved. Consequently, it can be argued that it underestimates the share of the emissions allocated to the electrical product.

4.2. Allocation Based on Exergy Content of Products

Allocations are evaluated in proportion to the exergy contents of the products, as follows:

$$f_E = Ex_E/(Ex_E + Ex_Q)$$

$$f_Q = Ex_Q/(Ex_E + Ex_Q)$$

where Ex_E and Ex_Q denote respectively the net outputs of electrical exergy and thermal exergy from the cogeneration system. It is noted that electrical energy and electrical exergy are equivalent, so that $Ex_E = E$. In this allocation method, one can treat the thermal product as simply thermal energy or thermal energy transfer via moving materials.

For simple thermal energy, the corresponding thermal exergy can be written as

$$ExQ = Q\tau$$

where τ denotes the exergetic temperature factor and can be evaluated as

$$\tau = 1 - T_o/T$$

Here, T denotes the temperature at which heat Q crosses the system boundary, and T_o denotes the temperature of the reference environment. For a reference environment at a temperature of 300 K (27°C), the value of the exergetic temperature factor τ is 0.25 for heat transfers at 400 K, 0.5 for heat transfers at 600 K, and 0.75 for heat transfers at 1200 K. For thermal energy is delivered via a material flow into and out of a heat exchange device, then the term Ex_Q is evaluated as the difference between the exergy of the incoming flow and the outgoing flow. The flowing commodity could be steam, hot water, cold water, etc., and the factors that must be taken into account in determining the corresponding exergy flow rates include mass flow rate, temperature, pressure and sometimes vapour fraction.

An additional point regarding the exergy-based allocation method is that the choice of reference environment for determining exergy quantities is important, and can affect the results. Normal practice involves selecting a reference environment that is similar to the actual environment. But other reference environments can be used. For example, Upton (2001), in an exercise to allocate emissions, evaluates exergy values using a reference-environment temperature of 100°C, which is well in excess of the actual environment annual mean temperature (perhaps 10°C to 20°C, depending on location). He uses this value because it is related to the process, in that he considers thermal energy below this temperature to be non-useful. Although this choice of a reference-environment temperature is permitted when using exergy methods, it is important to note that care must be exercised to ensure that a consistent reference environment is used throughout an analysis.

It is noted that the type of exergy-based allocation method presented by Strickland and Nyboer (2002a, 2002b) is not general, in that it assumes that the thermal product of cogeneration can be modelled only as pure thermal energy, rather than also considering the transfer of heat via flowing materials.

This allocation method accounts for the quality and quantity of the commodities involved. Consequently, compared to the energy allocation method, the exergy method avoids underestimating the share of the emissions allocated to the electrical product, and allocates a lower portion of the emissions to the thermal product.

4.3. Allocation Based on Economic Value of Products

Allocations are evaluated in proportion to the economic values of the products as follows:

$$f_E = c_E E/(c_E E + c_Q Q)$$

$$f_Q = c_E Q/(c_E E + c_Q Q)$$

where c_E and c_Q denote respectively the unit economic values of the electrical product and the thermal product of the cogeneration system. Note that the unit economic values presented here are based on energy (i.e., the economic value of a type of energy commodity per unit quantity of energy of that commodity), but the unit economic values based on other quantities—such as exergy—could also be used. Also, the economic values can be determined using several economic measures in several ways, e.g., they can reflect as the costs to produce the commodities, or their prices, or some other economic measure for them.

This method is sometimes considered to have advantages to owners of cogeneration systems that sell electrical and thermal products separately.

It is not necessary to know the numerical values of both unit economic values, c_E and c_Q, when applying this method of emission allocation. Rather, it is the ratio of the unit economic values that is important. This phenomenon can be seen by modifying the expressions for f_E and f_Q for this method of allocation as follows:

$$f_E = E/[E + Q(c_E/c_Q)^{-1}]$$

$$f_Q = Q/[E(c_E/c_Q) + Q]$$

Here, c_E/c_Q denotes ratio of the unit economic value for electricity to that for thermal energy. Since the unit economic value for electricity normally exceeds that for thermal energy, it is normally found that $c_E/c_Q > 1$.

4.4. Allocation Based on Incremental Fuel Consumption to Electrical Production

The emissions allocations are evaluated here by dividing the total fuel consumed in the cogeneration among the electrical and thermal products, while considering electricity generation to be a by-product of the thermal energy production process. Then, the emissions are allocated in proportion to the fuel division.

Two steps are used to divide the fuel consumption. First, the fuel consumption attributed to thermal energy production is evaluated as the hypothetical amount of fuel that would be consumed by an independent device for providing the same thermal energy as the cogeneration system (e.g., a reference steam boiler if the thermal energy is in the form of steam). That is,

$$F_Q = Q/\eta_b$$

where F_Q denotes the fuel consumption attributed to the production of thermal energy Q, and η_b denotes the energy efficiency of the independent device for providing the same thermal energy as the cogeneration system (e.g., a reference steam boiler). Second, the fuel consumption attributed to electricity generation, F_E, is evaluated by subtracting this hypothetical amount of fuel from the total primary fuel energy consumed by the cogeneration system, F. That is,

$$F_E = F - F_Q$$

Then, the emission allocation fractions f_E and f_Q are determined as the ratios F_E/F and F_Q/F, respectively. That is,

$$f_Q = Q/(F\eta_b)$$

$$f_E = 1 - f_Q$$

This allocation method is consistent with the "Fuel Charged to Power" (FCP) method used by many cogeneration consulting firms.

4.5. Allocation Based on Incremental Fuel Consumption to Thermal Energy Production

This emissions allocation method is similar to the previous one, except the emissions allocations are evaluated here by dividing the total fuel consumed in the cogeneration among the electrical and thermal products, while considering thermal energy production to be a by-product of the electricity generation process. Again, the emissions are then allocated in proportion to the fuel division, using the following two steps to divide the fuel consumption. First, the fuel consumption attributed to electricity generation is evaluated as the hypothetical amount of fuel that would be consumed by an independent device for providing the same electrical energy as the cogeneration system (e.g., a reference power plant). That is,

$$F_E = E/\eta_{pp}$$

where η_{pp} denotes the energy efficiency of the independent device for providing the same electrical energy as the cogeneration system (e.g., a reference power plant). Second, the fuel consumption attributed to thermal energy production, F_Q, is evaluated by subtracting this hypothetical amount of fuel from the total primary fuel energy consumed by the cogeneration system. That is,

$$F_Q = F - F_E$$

Again, the emission allocation fractions f_E and f_Q are then determined as the ratios F_E/F and F_Q/F, respectively, as follows:

$$f_E = E/(F\eta_{pp})$$

$$f_Q = 1 - f_E$$

4.6. Allocation Based on Shared Emission Savings between Electrical and Thermal Energy

The allocations are evaluated for each product in proportion to the hypothetical fuel that would be used to produce that product independently, relative to the total hypothetical fuel that would be used to produce both products independently. Using the terms previously defined, the hypothetical fuel consumption attributed to an independent process for thermal energy production (e.g., a reference steam boiler if the thermal energy is in the form of steam) is evaluated as

$$F_Q = Q/\eta_b$$

and the hypothetical fuel consumption attributed to an independent process for electricity generation is evaluated as

$$F_E = E/\eta_{pp}$$

Then, the emission allocation fractions f_E and f_Q are determined as the ratios F_E/F and F_Q/F, respectively, as follows:

$$f_E = (E/\eta_{pp})/(E/\eta_{pp} + Q/\eta_b)$$

$f_Q = (Q/\eta_b)/(E/\eta_{pp} + Q/\eta_b)$

This allocation method therefore shares the emissions among the products in a particular format. This method somewhat extends the concepts used in the previous two emissions allocation methods, but is more of a compromise in terms of treating one or the other product as the primary one.

4.7. Allocation by Agreement

Allocation of CO_2 emissions to each product of cogeneration can be determined purely based on an agreement between the various parties involved in a project.

5. Allocating Carbon Dioxide Emissions for Cogeneration

The author explains in this section why the most rational and meaningful method of allocating cogeneration CO_2 emissions is based on the exergy of the products. To support this contention, the different emissions allocation methods discussed for cogeneration processes in the previous section are examined and compared in this section. Before discussing and comparing the different emissions allocation methods, however, it is useful to understand the basic intentions and considerations in allocating emissions.

5.1. Basic Considerations in Allocating CO_2 Emissions for Cogeneration

The general objective when allocating a type of emission for a multi-product production process is to allocate the emission to each product according to the actual emission that is in fact attributable to that product, accounting for all thermodynamic losses, when it is produced in the multi-product production process. Usually the emission allocation breakdown is directly proportional to the breakdown of fuel use that is attributable to each product, when it is produced in the multi-product production process.

Following this description and considering carbon dioxide emissions for a cogeneration processes, the total CO_2 emissions C for a multi-product production process can be expressed as

$$C = C_E + C_Q$$

where C_E and C_Q denote respectively the CO_2 emissions associated with the electrical and thermal energy products, when they are produced in the cogeneration process. We can also express the total CO_2 emissions as

$$C = F\varphi$$

where F denotes the total fuel use in the process and φ a CO_2 emission coefficient for the fuel. The terms F and φ must be on consistent bases (e.g., if F is in energy units, then φ must be the CO_2 emission per unit fuel energy consumed). The total CO_2 emissions C can also be written as

$$C = (F_E + F_Q)\varphi$$

where F_E and F_Q denote respectively the fuel consumption associated with the electrical and thermal energy products, when they are produced in the cogeneration process. Clearly, $F_E + F_Q = F$. Furthermore, we can write

$$C = (f_E + f_Q)F\varphi$$

where

$$f_E = F_E/F$$

and

$$f_Q = F_Q/F$$

Clearly, the fractions of fuel consumption associated with the electrical and thermal energy products relate as follows: $f_E + f_Q = 1$.

5.2. Energy Considerations in Allocating Cogeneration CO_2 Emissions

Using an energy basis, the fuel consumption associated with generating the electricity in the cogeneration process can be expressed as

$$F_E = E/\eta_E$$

where η_E denotes the energy efficiency of generating the electrical energy product within a cogeneration process. Correspondingly, the fuel consumption associated with producing the thermal energy in the cogeneration process can be expressed as

$$F_Q = Q/\eta_Q$$

where η_Q denotes the energy efficiency of producing the thermal energy product within a cogeneration process. Combining the above equations, we can write the following expressions for the fractions f_E and f_Q:

$$f_E = (E/\eta_E)/(E/\eta_E + Q/\eta_Q)$$

and

$$f_Q = (Q/\eta_Q)/(E/\eta_E + Q/\eta_Q)$$

5.3. Exergy Considerations in Allocating Cogeneration CO_2 Emissions

Alternatively, we can use an exergy basis rather than an energy basis in establishing the above equations. Then, the fuel exergy consumption, Ex_{FE}, associated with generating the electrical exergy, Ex_E, in the cogeneration process can be expressed as

$$Ex_{FE} = Ex_E/\psi_E$$

where ψ_E denotes the exergy efficiency of generating the electrical energy product within the cogeneration process. Correspondingly, the fuel exergy consumption, Ex_{FQ}, associated with producing the thermal exergy, Ex_Q, in the cogeneration process can be expressed as

$$Ex_{FQ} = Ex_Q/\psi_Q$$

where ψ_Q denotes the exergy efficiency of producing the thermal energy product within the cogeneration process. Combining the above equations, we can write the following expressions for the fractions f_E and f_Q, using exergy terms:

$$f_E = (Ex_E/\psi_E)/(Ex_E/\psi_E + Ex_Q/\psi_Q)$$

and

$$f_Q /\psi_Q = (Ex_Q/\psi_Q)/(Ex_E/\psi_E + Ex_Q)$$

5.4. Advantages of Allocating Cogeneration CO_2 Emissions using Exergy

When an electrical generation process is modified so that it becomes a cogeneration process, some of the electrical product is sacrificed for a gain in thermal output. When considering energy quantities, it is usually seen that the thermal energy gain is often very great, even for a small decrease in electrical energy output. In addition, there is often no dependence on the temperature at which the thermal energy is delivered. When considering exergy quantities, however, the trade-off between electrical and thermal exergy products is more balanced. That is, a small decrease in electrical exergy output usually leads to a relatively small and similar in magnitude increase in thermal exergy output, while a large decrease in electrical exergy output usually leads to a correspondingly large increase in thermal exergy output. Furthermore, the increase in thermal exergy is directly dependent on the temperature at which the thermal energy is delivered; generally, the greater is the temperature the greater is the thermal exergy.

For example, a previous study (Rosen, 1990) of the effects of modifying a coal-fired electrical generating station for cogeneration showed that the overall variation in exergy efficiency is relatively small, while the corresponding variation for the energy efficiency is large. These results are illustrated in Table 2, where the exergy efficiencies are seen to vary between 35 and 39%, while the energy efficiencies vary between 37 and 69%. An interesting observation can be drawn. The exergy results demonstrate that the benefits of cogeneration are not really due to the shift from electricity generation to heat production, since there is a balanced trade-off between the exergy of the two product commodities, and the overall exergy efficiency remains relatively fixed. Rather, the benefits of cogeneration are due to the fact that the heat produced offsets the need for a separate heat production process that uses additional fuel and—on an exergy basis—is inefficient. The energy results present an entirely different perspective, one that is skewed due to the fact that energy analysis values electrical and thermal energy equally.

Table 2. Overall Energy and Exergy Efficiencies for an Electricity Generating Station Converted to Various Types of Cogeneration[1]

Operating mode	Temperature of product thermal energy (°C)	Energy efficiency (%)	Exergy efficiency (%)
Electricity generation only	–	37	37
Low-temperature cogeneration[2]	36	69	39
Intermediate-temperature cogeneration[2]	243	60	37
High-temperature cogeneration[2]	383	55	35

1. Based on data in Rosen (1990).
2. For cogeneration cases, 50% of the resulting process heat is assumed to be useful product.

The observation that a decrease in electrical exergy output of a cogeneration plant usually leads to a relatively similar magnitude increase in thermal exergy output, but that a decrease in electrical energy output of a cogeneration plant usually leads to a dissimilar magnitude increase in thermal energy output, suggests the following:

- The exergy efficiency of generating the electrical product within the cogeneration process, ψ_E, is similar to the exergy efficiency of generating the thermal product within the cogeneration process, ψ_Q.
- The energy efficiency of generating the electrical product within the cogeneration process, η_E, is not similar to the energy efficiency of generating the thermal product within the cogeneration process, η_Q.

As a consequence of the above two bullets, it can be seen from the analyses presented earlier that the allocation method based on exergy contents of the products most closely approximates the allocation expressions presented this section, and that the other emissions allocation methods in the prior section are significantly inaccurate relative to the objective of allocating emissions fairly and accurately. Thus, it can be seen that the exergy-based allocation method provides a rational means to determine the more productive modifications for a plant, when the objective is to reduce CO_2 emissions.

Another advantage of the method of allocating CO_2 emissions for cogeneration processes based on the exergy content of the products is that the allocation method is generalizable to any number and type of products. Most of the other allocation methods described earlier are much less flexible. For instance, the exergy-based method can accommodate trigeneration processes (i.e., cogeneration with cooling as a third product), and other processes producing two or more products, e.g., a fuel production process to produce hydrogen which also yields pure oxygen as a product or by-product, or a chemical process yielding two different chemical commodities.

6. Comparison of CO_2 Allocation Methods for Cogeneration

Based on the results of the previous section, the author proposes that the most rational and meaningful method of allocating carbon dioxide emissions for cogeneration processes is based on exergy. To justify this view, the different emissions allocation methods discussed for cogeneration processes in section 4 are compared. In particular, the problems inherent in the other CO_2 emission allocation methods for cogeneration processes are discussed.

The allocation method based on energy contents leads to inaccurate breakdowns of the carbon dioxide emissions, essentially because such a method presumes that the energy efficiency of generating the electrical product within a cogeneration process, η_E, is approximately similar to the energy efficiency of generating the thermal product within the cogeneration process, η_Q. As discussed in the last section, this presumption is not valid, as values for η_E and η_Q can vary widely.

The allocation method based on shared emission savings between electrical and thermal energy leads to inaccurate breakdowns of the carbon dioxide emissions, essentially because such a method presumes that the energy efficiency of generating the electrical product within a cogeneration process, η_E, is approximately similar to the energy efficiency of generating the electrical product via a separate process, η_{pp}. This presumption is not valid, as values for η_E and η_{pp} normally vary widely. Similarly, this allocation method presumes that the energy efficiency of generating the thermal product within a cogeneration process, η_Q, is approximately similar to the energy efficiency of generating the thermal product via a separate process, η_b, again a presumption that is not valid, as values for η_Q and η_b normally vary widely. It makes sense that these efficiencies vary since one of the main reasons to consider cogeneration is that it allows one to generate two products simultaneously with a higher efficiency than would be the case if each product were produced in a separate and independent process.

Note that one could determine the shared-emissions allocations based on exergy, rather than energy. Doing so would overcome many of the problems associated with the shared-emissions allocation method based on energy. This observation is attributable to the fact that the exergy-based efficiencies for electricity generation in the part of a cogeneration system responsible for electricity generation and in a pure electricity generation process are similar (i.e., $\psi_E \approx \psi_{pp}$), while the exergy-based efficiencies for thermal energy production in the part of a cogeneration system responsible for thermal energy production and in a pure thermal energy production process are similar (i.e., $\psi_Q \approx \psi_b$). Thus, the shared-emissions allocation method based on exergy reduces approximately to the allocation method based on the exergy contents of the products.

The allocation methods based on incremental fuel consumption to either electrical production or to thermal energy production both lead to inaccurate breakdowns of the carbon dioxide emissions. The reasons are similar and follow below:

- In essence, the allocation method based on incremental fuel consumption to electrical production presumes erroneously that the energy efficiency of generating the thermal product within a cogeneration process, η_Q, is approximately similar to the energy efficiency of generating the thermal product via a separate process, η_b. Still worse, the method then presumes that the value of the energy efficiency for generating the electrical product within the cogeneration process, η_E, can simply be selected so that the overall emissions total correctly. The ensuing values of η_E can as a result vary radically and for the most part arbitrarily.
- Similarly, the allocation method based on incremental fuel consumption to thermal energy production essentially presumes erroneously that the energy efficiency of generating the electrical product within a cogeneration process, η_E, is approximately similar to the energy efficiency of generating the electrical product via a separate process, η_{pp}. Further, the method then presumes that the value of the energy efficiency for generating the thermal product within the cogeneration process, η_Q, can simply be selected so that the overall emissions total correctly. As for the values of η_E in the preceding bullet, the ensuing values of η_Q can as a result vary radically and for the most part arbitrarily.

In general, the effect of the incremental-based allocations is that they arbitrarily underestimate the emissions from one of the products of a cogeneration process at the expense of the other. Both incremental-based allocations methods are thus unfair, since we seek the true and fair distribution of emissions among products—based on the efficiency of production for each within the cogeneration process.

It is noted that one could determine the incremental-based allocations based on exergy. However, this determination is not carried out here since the incremental allocation method is itself somewhat arbitrary and therefore not rational.

A common problem shared by the two incremental-based allocation methods and the shared-emissions allocation method is that they introduce independent devices for providing thermal energy (e.g., a reference steam boiler) and electrical energy (e.g., a reference power plant). The results obtained using these allocation methods are dependent on the energy efficiencies of these independent devices (η_b for the reference steam boiler and η_{pp} for the reference power plant). But, the values of η_b and η_{pp} can vary relatively widely depending on the specific devices chosen (e.g., high- versus medium- versus low-efficiency models), and these variations cause the emissions allocations evaluated with these methods to vary over correspondingly wide ranges.

The allocation method based on economic values of products leads to inaccurate breakdowns of the carbon dioxide emissions because that method allows economic parameters to skew the allocations. The proper allocations of carbon dioxide emissions for a cogeneration process should be based entirely on principles of thermodynamics or physics. Economic parameters such as costs and prices vary with time and location, but proper emissions allocations do not as they are dependent on characteristics of the technology involved. If one nevertheless chooses to modify the appropriate emissions allocations by penalizing certain products in terms of their emissions, through economic or other means, then it must be recognized that the resulting emissions allocations deviate arbitrarily from the appropriate emissions allocations.

Similarly, the allocation method based an agreement between the various stakeholders in a project leads to inaccurate breakdowns of the carbon dioxide emissions because that method allows arbitrary factors that generally are not based entirely on principles of thermodynamics or physics to affect the allocations. If one nevertheless chooses to modify the appropriate emissions allocations by penalizing certain products in terms of their emissions, through factors such as agreements between various stakeholders, then it must be recognized that the resulting emissions allocations almost certainly deviate arbitrarily from the appropriate emissions allocations.

Each allocation method in section 4, except that based on exergy, assign some arbitrary and/or subjective values to the differences between the product commodities. We need, instead, a rigorous scientific method, to help get the correct allocation and to remove the arbitrariness, and the exergy approach provides such a method. If, after determining the exergy-based allocations of CO_2 emissions, we nevertheless choose to allocate emissions differently—for economic, political or other reasons—we can do so, but at least we do so knowing the appropriate unbiased allocation.

7. Illustration 1: CO_2 Emissions Allocations for Actual Plants

The different methods for allocating carbon dioxide emissions for cogeneration plants are illustrated for two actual plants: the University of Toronto Cogeneration System and the Cornwall Cogeneration and District Energy System. The examples are intended to demonstrate the types of values that can be obtained using the different methods, and to illustrate the attributes of the different methods.

7.1. Process Descriptions and Data

The basic technical parameters for the two cogeneration systems considered, including data on the cogeneration engine and the cogenerated heat, are summarized in Table 3. Some general thermodynamic parameters for the cogeneration systems considered that are specified in the literature are presented in the top part of Table 4.

Energy and exergy data for the cogeneration systems considered are presented in the bottom two parts of Table 4. Those data include efficiencies and flow rates for products and inputs, and are based on data in the literature and evaluations by the present author. The energy and exergy flow rates for the fuel are similar, as are the energy and exergy flow rates for the electrical products. However, the product thermal energy and thermal exergy rates differ markedly for both processes, as do the energy and exergy efficiencies.

Table 3. Technical Parameters for the Cogeneration Systems Considered

Parameter	University of Toronto Cogeneration System[1]	Cornwall Cogeneration and District Energy System[2]
Engine type	Gas turbine	Two reciprocating engines
Heat use	Heating of campus using 6 km steam tunnel system	Municipal district heating using 4.5 km hot-water network
Heat quantity and type (base load)	30,000 lb/hr of 200 psi steam	7 MW via 120°C and 1585 kPa steam
Supplemental firing	Heat recovery steam generator can be supplementary-fired to 90,000 lb/hr steam at 200 psi	None
Environmental controls	Water injection to control nitrogen oxide emissions (to 42 ppm)	Engines use lean-burn technology
Installation date	1993	1995

1. Source is Consumers Gas (1995).
2. Source is Wiggin (1997).

7.2. Results and Discussion

The results of applying the methods for allocating CO_2 of emissions for the University of Toronto cogeneration plant are presented in Table 5 and Figure 4, and for the Cornwall Cogeneration and District Energy System are presented in Table 6 and Figure 5.

For both example cases, it is clear that the allocations of CO_2 emissions vary markedly, depending on the allocation method used. The author contends, as discussed throughout this article, that the exergy-based allocations are the most appropriate. Thus, using the other emissions allocation methods can be very misleading, since the resulting emissions may deviate widely from those obtained using the exergy-based method.

Some of the problems with the other allocation methods are illustrated in Tables 5 and 6 and Figures 4 and 5. Some examples follow:

- In one case (allocation based on incremental fuel consumption to heat production, for the Cornwall system), the absurd situation exists in which the emissions allocations to the thermal product are evaluated to be negative and those to the electrical product exceed 100%. This result is simply due to flaws in that allocation method and its use of an energy efficiency η_{pp} for an independent device for providing the same electrical energy as the cogeneration system (e.g., a reference power plant). Here, a value of η_{pp} = 35% is used. If, instead, the value of η_{pp} is selected such that η_{pp} = 38%, then the allocations of CO_2 emissions to both products would be positive and less than 100%.

- For the allocation method based on economic value of products, the results depend on the value of the ratio of the economic value of the electricity produced c_E to the economic value of the thermal energy produced c_Q. The value of this ratio, even considering only the present time, varies with location. In Tables 5 and 6, therefore, the values of the emission allocations for this method are left variable. In Figures 4 and 5, a range of c_E/c_Q values are considered.

Table 4. Specified and Evaluated Parameters for the Cogeneration Systems[1]

Parameter	University of Toronto Cogeneration Plant	Cornwall Cogeneration and District Energy System
General Thermodynamic Parameters		
Fuel type	Natural gas	Natural gas
Fuel input rate (kg/s)	0.3949	0.2660
Thermal-product type	Steam (dry saturated)	Hot water
Thermal-product temperature (°C)	197.6	120
Thermal-product absolute pressure (bar)	14.8	16.85
Thermal-product flow rate (kg/s)	3.78	15.12
Energy Parameters		
Fuel energy input rate (MW)	19.75	13.3
Electrical energy generation rate (MW)	6	5
Product thermal energy rate (MW)	10.393	7
Energy efficiency (LHV based) (%)	83	90
Exergy Parameters		
Fuel exergy input rate (MW)	20.35	13.70
Electrical exergy generation rate (MW)	6	5
Product thermal exergy rate (MW)	3.654	1.134
Exergy efficiency (%)	46.0	44.8

1. Reference-environment temperature and pressure are 10°C and 1 bar, respectively.

Table 5. Allocation of Emissions for University of Toronto Cogeneration Plant[1]

Emission-allocation method	Emission allocation (%)	
	To electrical product	To thermal product
Based on exergy content of products	62.1	37.9
Based on energy content of products	36.6	63.4
Allocation of incremental fuel consumption to electrical production[2]	41.5	58.5
Allocation of incremental fuel consumption to heat production[3]	86.8	13.2
Based on a shared emission savings between electricity and heat[2,3]	59.8	40.2
Based on economic value of products[4]	$5/[5 + 7(c_E/c_Q)^{-1}] \times 100\%$	$7/(5c_E/c_Q + 7) \times 100\%$

1. Reference-environment temperature and pressure are $T_o = 10°C$ and $p_o = 1$ bar, respectively.
2. An efficiency of 90% is assumed for the boiler that would have been used in the production of the same amount of heat as produced by the cogeneration system.
3. An efficiency of 35% is assumed for the power plant that would have been used in the production of the same amount of electricity as produced by the cogeneration system.
4. The parameter c_E/c_Q denotes the ratio of the economic value of the electricity produced to the economic value of the thermal energy produced.

Table 6. Allocation of Emissions for the Cornwall Cogeneration and District Heating System[1]

Emission-allocation method	Emission allocation (%)	
	To electrical product	To thermal product
Based on exergy content of products	81.5	18.5
Based on energy content of products	41.7	58.3
Allocation of incremental fuel consumption to electrical production[2]	41.5	58.5
Allocation of incremental fuel consumption to heat production[3]	107.4	−7.4
Based on a shared emission savings between electricity and heat[2,3]	64.7	35.3
Based on economic value of products[4]	$6/[6 + 10.4\,(c_E/c_Q)^{-1}] \times 100\%$	$10.4/(6c_E/c_Q + 10.4) \times 100\%$

1. Reference-environment temperature and pressure are $T_o = 10°C$ and $p_o = 1$ bar, respectively.
2. An efficiency of 90% is assumed for the boiler that would have been used in the production of the same amount of heat as produced by the cogeneration system.
3. An efficiency of 35% is assumed for the power plant that would have been used in the production of the same amount of electricity as produced by the cogeneration system.
4. The parameter c_E/c_Q denotes the ratio of the economic value of the electricity produced to the economic value of the thermal energy produced.

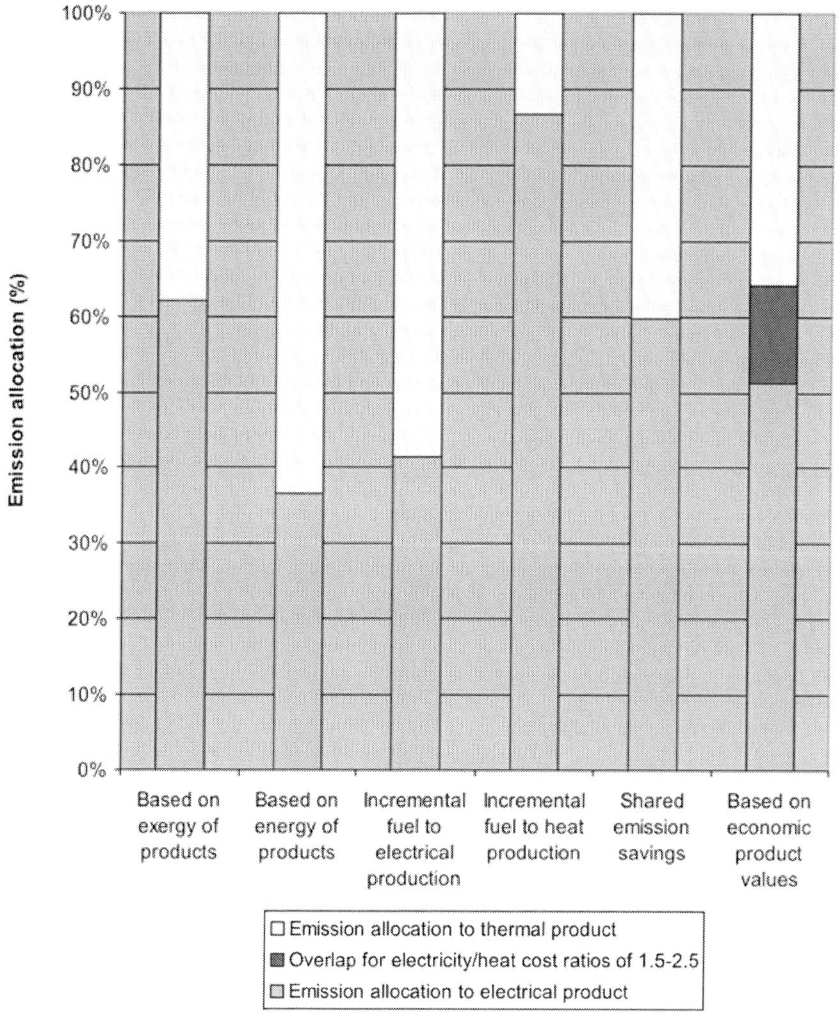

Figure 4. Allocation of emissions for University of Toronto cogeneration plant (based on data in Table 5).

8. Illustration 2: Comparison of CO_2 Emissions Allocations for a Cogeneration Plant and Equivalent Independent Plants

In this illustration, a hypothetical cogeneration scenario is compared to equivalent separate processes for producing thermal and electrical energy, and the implications of the different allocation methods are discussed. In addition, this case is used to illustrate how the results could be used to determine carbon dioxide emission credits that might be used in emissions trading.

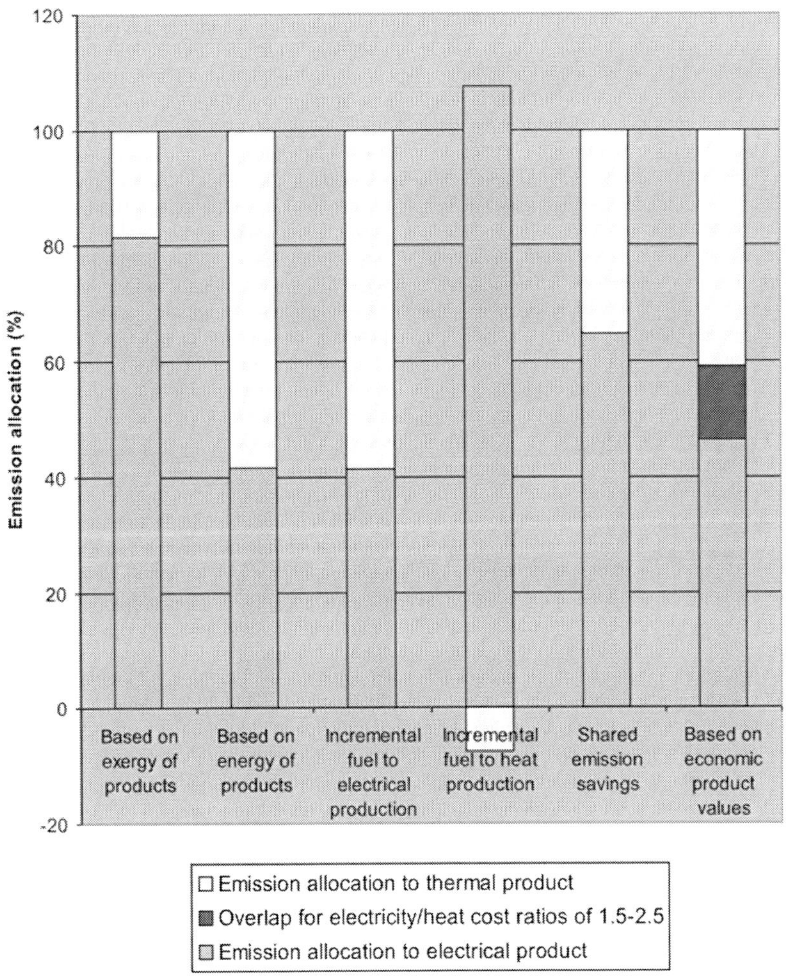

Figure 5. Allocation of emissions for Cornwall cogeneration and district heating system (based on data in Table 6).

8.1. Process Descriptions and Data

The main characteristics of the processes being compared are as follows:

- The hypothetical cogeneration system produces 4 MW of electrical power and 4 MW of thermal power from a fuel energy input rate of 10 MW. The energy efficiency is 80%.
- The separate processes consist of (i) an electricity generation system which produces 4 MW of electrical power from a fuel energy input rate of 10 MW, and (ii) a heating system which produces 4 MW of thermal power from a fuel energy input rate of 5 MW. The energy efficiency of the overall (combined) process is 53% (i.e., 8/15 × 100%).

In both cases, the input fuel is natural gas. The carbon dioxide emissions for natural gas are taken to be 50 kg CO_2/GJ natural gas.

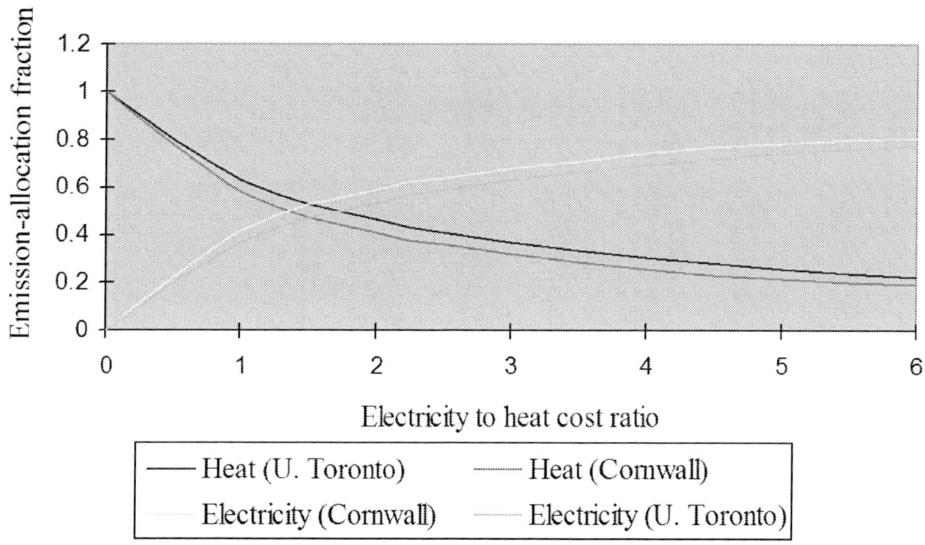

Figure 6. CO_2 emissions allocations based on economic values of products for two cases.

The 4 MW of thermal energy produced in each case is taken to be made up of 2 MW of steam and 2 MW of hot water. For simplicity, the state of the steam is taken to be the same as for the steam produced in the University of Toronto Cogeneration Plant described in the previous section (dry saturated steam at a pressure of 200 psi gauge), while the state of the hot water is taken to be the same as for the hot water produced in the Cornwall Cogeneration and District Energy System described in the previous section (hot water at a temperature of 120°C and a pressure of 1585 kPa gauge).

Energy and exergy data for the cogeneration and independent processes for heating and electricity generation are presented in Table 7. Those data include efficiencies and flow rates for products and inputs, and are based on evaluations by the present author. The data in Table 7 demonstrate that the energy and exergy flow rates for the fuel are similar, as are the energy and exergy flow rates for the electrical products. However, the product thermal energy and thermal exergy rates differ markedly for both processes, as do the energy and exergy efficiencies.

Two particular findings are observed in Table 7 regarding efficiency:

- Cogeneration is much more efficient (on energy or exergy bases) than the independent processes for producing the same thermal and electrical products.
- The exergy efficiencies are lower than the energy efficiencies, reflecting the fact that the thermal energy products are both of lower usefulness (or quality) than electricity. Of the two thermal products, the exergy values indicate that the usefulness of the steam is greater than that for the hot water

Table 7. Specified and Evaluated Thermodynamic Parameters for the Cogeneration and Independent Processes Considered[1]

Parameter	Independent processes			Cogeneration
	Heating	Electricity generation	Overall (combined)	
General Thermodynamic Parameters				
Fuel type	Natural gas	Natural gas	Natural gas	Natural gas
Fuel energy input rate (MW)	5	10	15	10
Thermal-product type(s)	Steam (dry sat.) Hot water	–	Steam (dry sat.) Hot water	Steam (dry sat.) Hot water
Thermal-product temperature (°C)	197.6 (steam) 120 (hot water)	–	197.6 (steam) 120 (hot water)	197.6 (steam) 120 (hot water)
Thermal-product absolute pressure (bar)	14.8 (steam) 16.85 (hot water)	–	14.8 (steam) 16.85 (hot water)	14.8 (steam) 16.85 (hot water)
Thermal-product flow rate (kg/s)	0.727 (steam) 4.32 (hot water)	–	0.727 (steam) 4.32 (hot water)	0.727 (steam) 4.32 (hot water)
Energy Parameters				
Fuel energy input rate (MW)	5	10	15	10
Electrical energy generation rate (MW)	–	4	4	4
Product thermal energy rate (MW)				
Steam	2	–	2	2
Hot water	2	–	2	2
Total	4	–	4	4
Energy efficiency (%)	80	40	53.3	80
Exergy Parameters				
Fuel exergy input rate (MW)	5.15	10.3	15.45	10.3
Electrical exergy generation rate (MW)	–	4	4	4
Product thermal exergy rate (MW)				
Steam	0.777	–	0.777	0.777
Hot water	0.323	–	0.323	0.323
Total	1.100	–	1.100	1.100
Exergy efficiency (%)	21.4	38.8	33.0	49.5

1. Reference-environment temperature and pressure are 10°C and 1 bar, respectively.

Table 8. Allocation (in %) of Emissions for the Cogeneration and Independent Processes Considered[1]

Process	Emission-allocation method	Emission allocation (%)	
		To electrical product	To thermal product
Independent			
Heating		0	100
Electricity generation		100	0
Overall (combined)		66.7	33.3
Cogeneration	Based on exergy content of products	78.4	21.6
	Based on energy content of products	50.0	50.0
	Allocation of incremental fuel consumption to electrical production[2]	55.6	44.4
	Allocation of incremental fuel consumption to heat production[3]		
	Assuming a reference power plant efficiency of $\eta_{pp} = 35\%$	114.3	–14.3
	Assuming a reference power plant efficiency of $\eta_{pp} = 40\%$	100	0
	Based on a shared emission savings between electricity and heat[2,3]		
	Assuming a reference power plant efficiency of $\eta_{pp} = 35\%$	72.0	28.0
	Assuming a reference power plant efficiency of $\eta_{pp} = 40\%$	69.2	30.8
	Based on economic value of products[4]		
	Assuming an electrical-to-thermal cost ratio of $c_E/c_Q = 1.5$	60.0	40.0
	Assuming an electrical-to-thermal cost ratio of $c_E/c_Q = 1.8$	64.3	35.7
	Assuming an electrical-to-thermal cost ratio of $c_E/c_Q = 2.1$	67.7	32.3

1. Reference-environment temperature and pressure are $T_o = 10°C$ and $p_o = 1$ bar, respectively.
2. An efficiency of 90% is assumed for the boiler that would have been used in the production of the same amount of heat as produced by the cogeneration system.
3. Efficiencies of 35% and 40% are considered for the power plant that would have been used in the production of the same amount of electricity as produced by the cogeneration system.
4. The parameter c_E/c_Q denotes the ratio of the economic value of the electricity produced to the economic value of the thermal energy produced.

Table 9. Allocation (in g CO_2/s) of Emissions for the Cogeneration and Independent Processes Considered[1]

Process	Emission-allocation method	Emission allocation ($g\ CO_2/s$)		
		To electrical product	To thermal product	Total
Independent				
Heating		0	250	250
Electricity generation		500	0	500
Overall (combined)		500	250	750
Cogeneration	Based on exergy content of products	392	108	500
	Based on energy content of products	250	250	500
	Allocation of incremental fuel consumption to electrical production[2]	278	222	500
	Allocation of incremental fuel consumption to heat production[3]			
	Assuming a reference power plant efficiency of η_{pp} = 35%	572	–72	500
	Assuming a reference power plant efficiency of η_{pp} = 40%	500	0	500
	Based on a shared emission savings between electricity and heat[2,3]			
	Assuming a reference power plant efficiency of η_{pp} = 35%	360	140	500
	Assuming a reference power plant efficiency of η_{pp} = 40%	346	154	500
	Based on economic value of products[4]			
	Assuming an electrical-to-thermal cost ratio of $c_E/c_Q = 1.5$	300	200	500
	Assuming an electrical-to-thermal cost ratio of $c_E/c_Q = 1.8$	322	178	500
	Assuming an electrical-to-thermal cost ratio of $c_E/c_Q = 2.1$	339	161	500

1. Reference-environment temperature and pressure are $T_o = 10°C$ and $p_o = 1$ bar, respectively.
2. An efficiency of 90% is assumed for the boiler that would have been used in the production of the same amount of heat as produced by the cogeneration system.
3. Efficiencies of 35% and 40% are considered for the power plant that would have been used in the production of the same amount of electricity as produced by the cogeneration system.
4. The parameter c_E/c_Q denotes the ratio of the economic value of the electricity produced to the economic value of the thermal energy produced

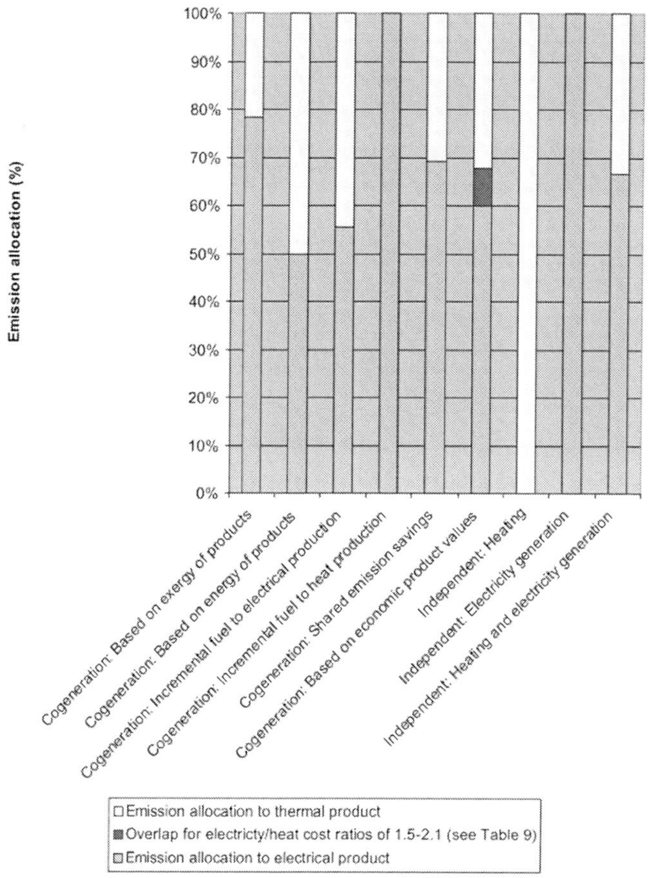

Figure 7. Allocation of emissions (in %) for independent and cogeneration processes considered (based on data in Table 8 and a reference power plant efficiency of 40%).

8.2. Results and Discussion

The results of applying the methods for allocating CO_2 of emissions for the cogeneration and independent processes are presented in Tables 8 and 9 and Figures 7 and 8. Percentage breakdowns are shown in Table 8 and Figure 7, while absolute emissions rates are shown in Table 9 and Figure 8.

For the independent processes, there exists no ambiguity regarding the allocations of emissions to the thermal and electrical products. For the cogeneration process, the appropriate method to allocate emissions is not clear, so the allocations methods discussed earlier are applied. It is observed that the allocations of CO_2 emissions vary markedly, depending on the allocation method used.

The author contends, as discussed throughout this article, that the exergy-based allocations are the most appropriate. Thus, using the other emissions allocation methods can be very misleading, since the resulting emissions may deviate widely from those obtained using the exergy-based method.

Some of the problems with the other allocation methods are illustrated in Tables 8 and 9 and Figures 7 and 8. Some examples follow:

- In one case (where the allocation is based on incremental fuel consumption to heat production), the absurd situation exists in which the allocations of emissions to the thermal product are evaluated to be negative and the allocations to the electrical product to exceed 100%. This result is simply a consequence of the flaws in that allocation method and its use of an energy efficiency η_{pp} for an independent device for providing the same electrical energy as the cogeneration system (e.g., a reference power plant). Here, a value of η_{pp} = 35% is used. For comparison, an alternate value of η_{pp} is selected (40%); then the allocations of carbon dioxide emissions to both products do not exceed 100%.

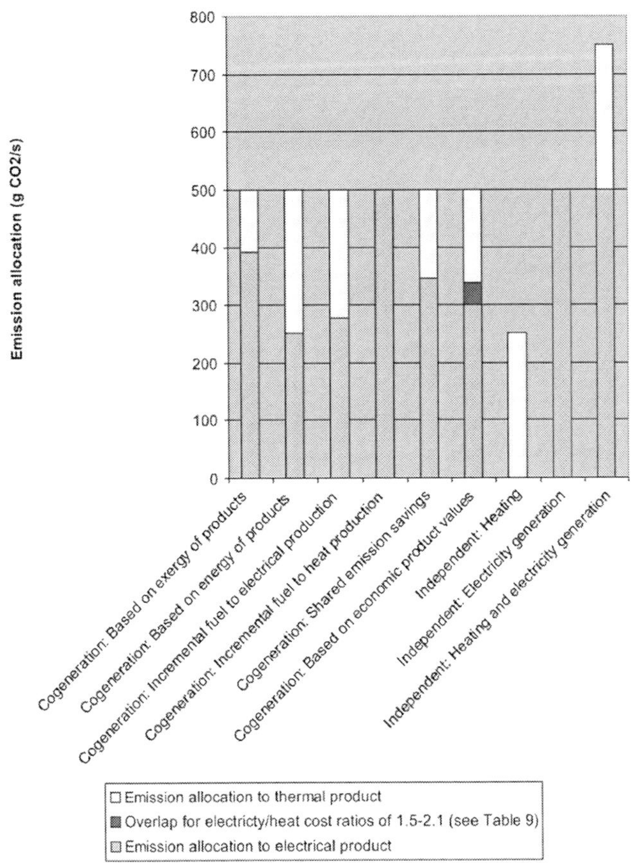

Figure 8. Allocation of emissions (in g CO_2/s) for independent and cogeneration processes (based on data in Table 9 and a reference power plant efficiency of 40%).

- For the allocation method based on economic value of products, the results depend on the value of the ratio of the economic value of the electricity produced c_E to the economic value of the thermal energy produced c_Q. The value of this ratio, even considering only the present time, varies with location. Here, for illustration only, economic parameter values from a recent U.S. report (Harrell, 2002) are used. With these values, the cost of electricity is approximated as US\$0.060/kWh, and the cost of steam as US\$0.0334. The electrical-to-thermal cost ratio is then $c_E/c_Q = 1.8$. In Tables 8 and 9, therefore, this value is used. Also, values of c_E/c_Q of 1.5 and 2.1 are used to indicate the sensitivity of the allocations to the cost ratio.

9. Determining CO_2 Emissions Credits for Trading Purposes from Switching to Cogeneration

In this section, the case in section 8 is used to illustrate how the results could be used to determine carbon dioxide emission credits that might be used in emissions trading. Here, we consider two energy users, one of electricity and one of thermal energy. The types of thermal energy used are the same as those described in section 8. The decrease in CO_2 emissions attributable to the energy users are evaluated, when each switches from obtaining the energy required (electricity or thermal energy) from a producer of the just the required energy to a supplier using cogeneration. The decrease in CO_2 emissions for each energy user is its CO_2 emissions credits. The characteristics of the cogeneration plant and the independent electrical power plant and heating plant are as in section 8.

9.1. CO_2 Emissions Credits when Electricity User Converts to Cogeneration

In this situation, we consider an electrical consumer who normally obtains electricity from a power plant. We wish to determine the decrease in CO_2 emissions attributable to that consumer (i.e., the CO_2 emissions credit for the consumer) if the consumer switches to obtaining electricity from a supplier that uses a cogeneration plant.

We consider a multi-step calculation procedure. For simplicity, we consider a unit energy use by the consumer of 1 GJ of electricity. Parameter values that characterize the technologies considered are drawn from section 8.

The CO_2 emissions attributable to the consumer when obtaining electricity from a power plant can be evaluated as the product of the fuel use in the power plant and the carbon dioxide emissions factor for the fuel. That is,

CO_2 emissions = (Fuel use in power plant)(Fuel CO_2 emissions factor)
= (Electricity use/Power plant efficiency)(Fuel CO_2 emissions factor)
= (1 GJ electricity/0.40 GJ electricity/GJ fuel)(50 kg CO_2 /GJ fuel)
= 125.0 kg CO_2 /GJ electricity

The total CO_2 emissions of the cogeneration plant (to produce 1 GJ of electricity as well as thermal energy) can be evaluated as the product of the total fuel use in the cogeneration plant and the carbon dioxide emissions factor for the fuel. That is,

CO_2 emissions = (Fuel use in cogeneration plant)
(Fuel CO_2 emissions factor)

= (Electricity use/Cogen. efficiency for elec.)(Fuel CO_2 emissions factor)
= (1 GJ electricity/0.40 GJ electricity/GJ fuel)(50 kg CO_2 /GJ fuel)
= 125.0 kg CO_2 /GJ electricity

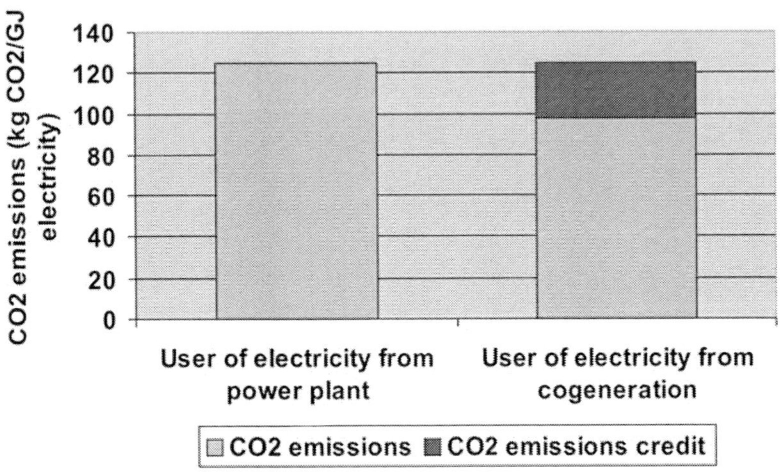

Figure 9. Carbon dioxide emissions for electricity generation via power plant and cogeneration, and carbon dioxide emissions credit.

Of the total CO_2 emissions of the cogeneration plant (to produce 1 GJ of electricity as well as thermal energy), the CO_2 emissions attributable to the consumer when obtaining electricity from a cogeneration plant can be evaluated as the fraction of the total CO_2 emissions of the cogeneration plant attributed to electricity production. In this article, we have argued that the division of CO_2 emissions for cogeneration among electrical and thermal products should be based on the exergy contents of the products. For the present case, it was shown in section 8 that 78.4% of the total CO_2 emissions for the cogeneration plant should be attributed to the electrical product based on exergy (see Table 8). Thus, the CO_2 emissions attributable to the consumer when obtaining electricity from a cogeneration plant can be evaluated as follows:

CO_2 emissions for user = (Total CO_2 emissions of cogen. plant)
(Fraction for elec.)
= (125.0 kg CO_2 /GJ electricity)(0.784)
= 98.0 kg CO_2 /GJ electricity

Finally, the CO_2 emissions credit for switching to cogeneration, evaluated as the decrease in CO_2 emissions attributable to the electricity user, can be evaluated as the difference between the CO_2 emissions attributable to the consumer when obtaining electricity from a power plant and from a cogeneration plant. That is,

CO_2 emissions credit = (CO_2 emissions for elec. from power plant
attributed to consumer)
– (CO_2 emissions for elec. from cogen. attributed to consumer)
= (125 kg CO_2 /GJ electricity) – (98.0 kg CO_2 /GJ electricity)
= 27.0 kg CO_2 /GJ electricity

The results are presented in Table 10 and illustrated in Figure 9, and demonstrate the manner in which they can assist in evaluating CO_2 emissions credits for electricity users, for trading and other purposes.

Table 10. Carbon Dioxide Emissions and Emissions Credit for Switching to Cogeneration for Electricity and Heat Users*

	CO_2 emissions (kg)
For User of Electricity (1 GJ)	
CO_2 emissions for user of electricity from power plant	125.0
CO_2 emissions for user of electricity from cogeneration plant	98.0
CO_2 emissions credit for user of electricity for switching from power plant to cogeneration plant	27.0
For User of Heat (1 GJ)	
CO_2 emissions for user of heat from heating plant	62.5
CO_2 emissions for user of heat from cogeneration plant	27.0
CO_2 emissions credit for user of heat for switching from heating plant to cogeneration plant	35.5

* Allocation of emissions from cogeneration to electrical and heat products is determined based on exergy contents of products.

9.2. CO_2 Emissions Credits when Heat User Converts to Cogeneration

In this situation, we consider a heat consumer who normally obtains heat from a heating plant. We wish to determine the decrease in CO_2 emissions attributable to that consumer (i.e., the CO_2 emissions credit for the consumer) if the consumer switches to obtaining heat from a supplier that uses a cogeneration plant.

We consider a multi-step calculation procedure. For simplicity, we consider a unit energy use by the consumer of 1 GJ of thermal energy. The thermal energy used by the consumer is of the type described in section 8. Parameter values that characterize the technologies considered are drawn from section 8

The CO_2 emissions attributable to the consumer when obtaining heat from a heating plant can be evaluated as the product of the fuel use in the heating plant and the carbon dioxide emissions factor for the fuel. That is,

CO_2 emissions = (Fuel use in heating plant)(Fuel CO_2 emissions factor)
= (Heat use/Heating plant efficiency)(Fuel CO_2 emissions factor)
= (1 GJ heat/0.80 GJ heat/GJ fuel)(50 kg CO_2 /GJ fuel)
= 62.5 kg CO_2 /GJ heat

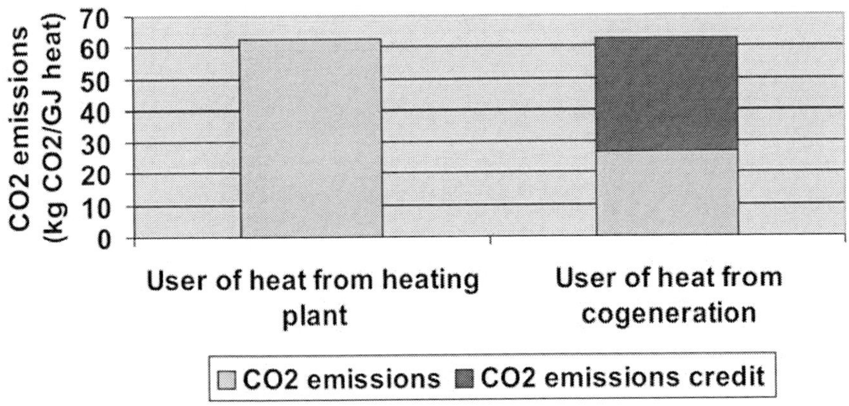

Figure 10. CO_2 emissions for heat production via heating plant and cogeneration, and CO_2 emissions credit.

The total CO_2 emissions of the cogeneration plant (to produce 1 GJ of heat as well as electrical energy) can be evaluated as the product of the total fuel use in the cogeneration plant and the carbon dioxide emissions factor for the fuel. That is,

CO_2 emissions = (Fuel use in cogeneration plant)(Fuel CO_2 emissions factor)
= (Electricity use/Cogen. efficiency for heat)(Fuel CO_2 emissions factor)
= (1 GJ heat/0.40 GJ heat/GJ fuel)(50 kg CO_2 /GJ fuel)
= 125.0 kg CO_2 /GJ heat

Of the total CO_2 emissions of the cogeneration plant (to produce 1 GJ of heat as well as electrical energy), the CO_2 emissions attributable to the consumer when obtaining heat from a cogeneration plant can be evaluated as the fraction of the total CO_2 emissions of the cogeneration plant attributed to heat production. In this article, we have argued that the division of CO_2 emissions for cogeneration among electrical and thermal products should be based on the exergy contents of the products. For the present case, it was shown in section 8 that 21.6% of the total CO_2 emissions for the cogeneration plant should be attributed to the thermal product based on exergy (see Table 8). Thus, the CO_2 emissions attributable to the consumer when obtaining heat from a cogeneration plant can be evaluated as follows:

CO_2 emissions for user = (Total CO_2 emissions of cogeneration plant)(Fraction for heat)
= (125.0 kg CO_2 /GJ heat)(0.216)
= 27.0 kg CO_2 /GJ heat

Finally, the CO_2 emissions credit for switching to cogeneration, evaluated as the decrease in CO_2 emissions attributable to the heat user, can be evaluated as the difference between the CO_2 emissions attributable to the consumer when obtaining heat from a power plant and from a cogeneration plant. That is,

CO_2 emissions credit =(CO_2 emissions for heat from heating plant attributed to consumer)
– (CO_2 emissions for heat of cogen. attributed to consumer)

$= (62.5 \text{ kg } CO_2/\text{GJ heat}) - (27.0 \text{ kg } CO_2/\text{GJ heat})$
$= 35.5 \text{ kg } CO_2/\text{GJ heat}$

The results are presented in Table 10 and illustrated in Figure 10, and demonstrate the manner in which they can assist in evaluating CO_2 emissions credits for thermal energy users, for trading and other purposes.

9.3. CO_2 Emissions Credits for Other Cases

The procedures illustrated in the previous two subsections can be formalized for variations of the cases considered here, as well as for various other cases. Some of the other situations that could be considered include on- and off-site plants, different fuels, different thermal energy requirements and trigeneration systems.

Conclusion

The exergy-based method of carbon dioxide emissions allocation allows for rational and meaningful allocations of such emissions for cogeneration, and is superior to other methods. By permitting carbon dioxide emissions to be allocated more appropriately among the different commodities generated in cogeneration, the results allow the environmental benefits of technologies that produce multiple products to be better understood and exploited. These results should therefore allow the more beneficial among competing technologies to be identified in a rational and meaningful manner. The results also indicate that the exergy-based emissions allocations method provides a sensible basis for a meaningful overall approach for emissions trading. The results suggest that the exergy-based method should be used in allocating carbon dioxide emissions for cogeneration devices, as using the exergy-based method would help ensure proper decision-making regarding issues such as

- how emissions should be reduced in a given device,
- how and where cogeneration technologies should be used,
- what effect introducing cogeneration will have on overall carbon dioxide emissions, and
- a fair way to establish detailed schemes for emissions trading.

To advance this approach, the exergy-based allocation method needs to be applied with more detailed process (e.g. hourly data for a plant over a year, rather than annual average data). Also, detailed methodologies for emissions trading should be developed that utilize overall exergy-based emissions allocations approach. Nonetheless, the results presented in this article should be of benefit to designers of energy systems, and to decision and policy makers in companies and government. If the results are used appropriately, they should allow benefits to accrue to society through the selection and design of better energy technologies, based on environmental considerations. The results may also be useful as input to the regulations developed in Europe for defining and assessing the benefits of cogeneration

(European Parliament, 2004; CEN/CENELEC, 2004). In particular, the incorporation of exergy into these regulations would likely improve them markedly, both in general and in particular for allocating emissions.

ACKNOWLEDGMENTS

The support provided by the Oil, Gas and Energy Branch of Environment Canada is gratefully acknowledged.

NOMENCLATURE

c_E	unit economic value of electrical product of cogeneration
c_Q	unit economic value of thermal product of cogeneration
C	total CO_2 emissions from cogeneration
C_E	CO_2 emissions for electrical energy produced via cogeneration
C_Q	CO_2 emissions for thermal energy product produced via cogeneration
E	net output of electrical energy from cogeneration
Ex_E	net output of electrical exergy from cogeneration
Ex_{FE}	fuel exergy consumption for generating electricity via cogeneration
Ex_{FQ}	fuel exergy consumption for producing thermal exergy via cogeneration
Ex_Q	net output of thermal exergy from cogeneration
f_E	fraction of cogeneration emissions allocated to electrical product
f_Q	fraction of cogeneration emissions allocated to thermal product
F	total primary fuel energy consumed by cogeneration system
F_E	fuel consumption attributed to electricity generation
F_Q	fuel consumption attributed to production of thermal energy
Q	net output of thermal energy from cogeneration
T	temperature
T_o	temperature of reference environment

GREEK SYMBOLS

φ	CO_2 emission coefficient for a fuel
η_b	energy efficiency of independent device (e.g., boiler) for thermal energy
η_E	energy efficiency of generating electrical energy via cogeneration
η_Q	energy efficiency of producing thermal energy via cogeneration
η_{pp}	energy efficiency of independent device (e.g., power plant) for electricity
ψ_E	exergy efficiency of generating electricity via cogeneration
ψ_Q	exergy efficiency of producing thermal energy product via cogeneration
τ	exergetic temperature factor

REFERENCES

Acres International Limited (1987). *Cogeneration Potential in Ontario and Barriers to its Development*, Report for Ontario Ministry of Energy, Toronto.

Ayres, R. U., Ayres, L. W. and Martinas, K. (1998). Exergy, waste accounting, and life-cycle analysis. *Energy* 23, 355-363.

Berthiaume, R., Bouchard, C. and Rosen, M. A. (2001). Exergetic evaluation of the renewability of a biofuel. *Exergy, An Int. J.* 1, 256-268.

CEN/CENELEC (2004). *Workshop Agreement Manual for Determination of Combined Heat and Power (CHP),* CWA 45547, European Committee for Standardization (CEN) and European Committee for Electrotechnical Standardization (CENELEC), Brussels.

Consumers Gas (1995, estimated). Cogeneration at the University of Toronto. Information sheet.

Connelly, L. and Koshland, C. P. (1997). Two aspects of consumption: using an exergy-based measure of degradation to advance the theory and implementation of industrial ecology. *Resources, Conservation and Recycling* 19, 199-217.

Cownden, R., Nahon, M. and Rosen, M. A. (2001). Exergy analysis of a fuel cell power system for transportation applications. *Exergy, An Int. J.* 1, 112-121.

Crane, P., Scott, D. S. and Rosen, M. A. (1992). Comparison of exergy of emissions from two energy conversion technologies, considering potential for environmental impact. *Int. J. Hydrogen Energy* 17, 345-350.

Creyts, J. C. and Carey, V. P. (1997). Use of extended exergy analysis as a tool for assessment of the environmental impact of industrial processes. *Proc. ASME Advanced Energy Systems Division*, ed. M.L. Ramalingam, ASME, New York, AES-Vol. 37, 129-137.

Daniel, J. J. and Rosen, M. A. (2002). Exergy-based environmental assessment of life cycle emissions for various automobiles and fuels. *Proc. CSME Forum* 2002, 21-24 May, Kingston, Ontario, Sec. 20, Paper 4, pp. 1-14.

Dincer, I. and Rosen, M. A. (2007). *Exergy: Energy, Environment and Sustainable Development*, Elsevier, Oxford, UK.

Edgerton, R. H. (1982). *Available Energy and Environmental Economics*, D.C. Heath, Toronto.

European Parliament (2004). Directive 2004/8/EC of the European Parliament and of the Council of 11 February 2004 on the Promotion of Cogeneration Based on a Useful Heat Demand in the Internal Energy Market and Amending Directive 92/42/EEC, *Official Journal of the European Union* (L52) 47, 50-60.

FVB/Eltec (1993). *Potential Heat Production from Existing and Future Electric Generating Plants*, Report No. 9102-G-806 for Canadian Electrical Association.

Gunnewiek, L. H. and Rosen, M. A. (1998). Relation between the exergy of waste emissions and measures of environmental impact. *Int. J. Environment and Pollution* 10, 261-272.

Harrell, G. (2002). *Steam System Survey Guide*, Report ORNL/TM-2001/263 for Oak Ridge National Laboratory.

Hart, D. R. and Rosen, M. A. (1994). Environmental and health benefits of utility-based cogeneration in Ontario, Canada. *Energy and Environment* 5, 363-378.

Henneforth, J. C. and Todd, D. M. (1988, August). Cogeneration on a large scale. *Mechanical Engineering* 110, 51-53.

Hevert, H. W. and Hevert, S. C. (1980). Second law analysis: an alternative indicator of system efficiency, *Energy-The International Journal* 5, 865-873.

Horlock, J. H. (1987). Some practical CHP schemes. Chapter 5 of *Cogeneration: Combined Heat and Power (CHP)*, Pergamon, Oxford, England, pp. 158-174.

Klein, M. (1999, estimated). *Brief Descriptions of Commercial and Institutional CHP Systems in Canada.* Report by Oil, Gas and Energy Branch, Environment Canada.

Kotas, T.J. (1995). *The Exergy Method of Thermal Plant Analysis*, reprint edition, Krieger, Malabar, Florida.

MacLaren Engineers Inc. (1988). *Cogeneration Sourcebook*, Report for Ontario Ministry of Energy, Toronto.

MacRae, K.M. (1992). *Realizing the Benefits of Community Integrated Energy Systems*, Canadian Energy Research Institute, Calgary, Alberta.

Moran, M.J. (1989). *Availability Analysis: A Guide to Efficient Energy Use*, Revised Ed., ASME, New York.

Moran, M.J. and Sciubba, E. (1994). Exergy analysis: principles and practice. *J. Engineering for Gas Turbines and Power* 116, 285-290.

Moran, M.J. and Shapiro, H.N. (2007). *Fundamentals of Engineering Thermodynamics*, 6th ed., Wiley, New York.

Phylipsen, G.J.M., Blok, K. and Worrell, E. (1998). *Handbook on International Comparisons of Energy Efficiency in the Manufacturing Industry.* Department of Science, Technology and Society, Utrecht University, the Netherlands.

Rogner, H.-H. (1993). Clean energy services without pain: district energy systems. *Energy Studies Review* 5, 114-120.

Rosen, M.A. (1990). Comparison based on energy and exergy analyses of the potential cogeneration efficiencies for fuel cells and other electricity generation devices. *Int. J. Hydrogen Energy* 15, 267-274.

Rosen, M.A. (1992). Evaluation of energy utilization efficiency in Canada using energy and exergy analyses. *Energy-The International Journal* 17, 339-350.

Rosen, M.A. (1993). Energy utilization efficiency in a macrosystem (Ontario): evaluation and improvement through cogeneration. *Proc. Int. Symp. CO_2 Fixation and Efficient Utilization of Energy*, 29 Nov.-1 Dec., Tokyo, pp. 17-26.

Rosen, M.A. (1994). Assessment of various scenarios for utility-based cogeneration in Ontario. *Energy-The International Journal* 19, 1143-1149.

Rosen, M.A. (1996). Thermodynamic investigation and comparison of selected production processes for hydrogen and hydrogen-derived fuels. *Energy-The International Journal* 21, 1079-1094.

Rosen, M.A. (1998). Reductions in energy use and environmental emissions achievable with utility-based cogeneration: simplified illustrations for Ontario. *Applied Energy* 61, 163-174.

Rosen, M.A. (1999). Second law analysis: approaches and implications. *Int. J. Energy Research* 23, 415-429.

Rosen, M.A. (2000). Thermodynamic comparison of coal-fired and nuclear electrical generating stations. *Trans. Can. Soc. Mech. Eng.* 24 (1B), 273-283.

Rosen, M.A. and Dincer, I. (1997). On exergy and environmental impact. *Int. J. Energy Research* 21, 643-654.

Rosen, M.A. and Dincer, I. (1999). Exergy analysis of waste emissions. *Int. J. Energy Research* 23, 1153-1163.

Rosen, M.A. and Dincer, I. (2001). Exergy as the confluence of energy, environment and sustainable development. *Exergy, An Int. J.* 1, 3-13.

Rosen, M.A. and Dincer, I. (2002). Energy and exergy analyses of thermal energy storage systems. Chapter 10 of *Thermal Energy Storage Systems and Applications*, Wiley, London, pp. 411-510.

Rosen, M.A. and Horazak, D.A. (1995). Energy and exergy analyses of PFBC power plants. Chapter 11 of *Pressurized Fluidized Bed Combustion*, ed. M. Alvarez Cuenca and E.J. Anthony, Chapman and Hall, London, England, pp. 419-448.

Rosen, M.A. and Le, M. (1994). Assessment of the potential cumulative benefits of applying utility-based cogeneration in Ontario. *Energy Studies Review* 6, 154-163.

Rosen, M.A. and Le, M. (1996). Efficiency analysis of a process design integrating cogeneration and district energy. *Proc. ASME Advanced Energy Systems Division*, AES-Vol. 36, ed. A.B. Duncan, J. Fiszdon, D. O'Neal and K. Den Braven, ASME, New York, pp. 473-480.

Rosen, M.A., Le, M.N. and Dimitriu, J. (1997). Efficiency improvement for large energy systems through utility-based cogeneration and district energy: investigation for Ontario, Canada. *World Energy System: An International Journal* 1, 56-63.

Sato, N. (2005). *Chemical Energy and Exergy: An Introduction to Chemical Thermodynamics for Engineers*, Elsevier, Oxford, UK.

Sciubba, E. (1999). Exergy as a direct measure of environmental impact. *Proc. ASME Advanced Energy Systems Division*, ed. S.M. Aceves et al., AES-Vol. 39, 573-581.

Sciubba, E. (2004). From engineering economics to extended exergy accounting: A possible path from monetary to resource-based costing. *Journal of Industrial Ecology* 8(4), 19-40.

Strickland, C. and Nyboer, J. (2002a). *Cogeneration Potential in Canada: Phase 2. Report for Natural Resources Canada*, by MK Jaccard and Associates.

Strickland, C. and Nyboer, J. (2002b). *A Review of Existing Cogeneration Facilities in Canada.* Report by Canadian Industrial Energy End-Use Data and Analysis Center, Simon Fraser University.

Szargut, J. (1980). International progress in second law analysis, *Energy* 5, 709-718.

Szargut, J. (2005). *Exergy Method: Technical and Ecological Applications*, WIT Press, Southampton, UK.

Szargut, J., Morris, D.R. and Steward, F.R. (1988). *Exergy Analysis of Thermal, Chemical and Metallurgical Processes*, Hemisphere, New York.

Upton, B. (2001). Letter of 16 November 2001 to P. Bhatia, World Resources Institute, Washington, DC, from B. Upton, Senior Research Engineer, National Council for Air and Stream Improvement, Inc., West Coast Regional Center, Corvallis, OR.

Verbruggen, A. (2007a). What's needed next to refine the EU directive on cogeneration regulation. *The Electricity Journal* 20(2), 63-70.

Verbruggen, A. (2007b). Quantifying combined heat and power (CHP) activity, *Int. J. Energy Tech. and Policy* 5(1), 17-35.

Wiggin, M. (1997, estimated). *Cornwall, Ontario District Heating System.* Information sheet, CANMET Energy Technology Centre, Natural Resources Canada.

INDEX

D

E

F

G

H

I

K

L

M

N

O

P

U

V

W

X

Y